Electronic Design Automation Frameworks

IFIP – The International Federation for Information Processing

IFIP was founded in 1960 under the auspices of UNESCO, following the First World Computer Congress held in Paris the previous year. An umbrella organization for societies working in information processing, IFIP's aim is two-fold: to support information processing within its member countries and to encourage technology transfer to developing nations. As its mission statement clearly states,

> IFIP's mission is to be the leading, truly international, apolitical organization which encourages and assists in the development, exploitation and application of information technology for the benefit of all people.

IFIP is a non-profitmaking organization, run almost solely by 2500 volunteers. It operates through a number of technical committees, which organize events and publications. IFIP's events range from an international congress to local seminars, but the most important are:

- the IFIP World Computer Congress, held every second year;
- open conferences;
- working conferences.

The flagship event is the IFIP World Computer Congress, at which both invited and contributed papers are presented. Contributed papers are rigorously refereed and the rejection rate is high.

As with the Congress, participation in the open conferences is open to all and papers may be invited or submitted. Again, submitted papers are stringently refereed.

The working conferences are structured differently. They are usually run by a working group and attendance is small and by invitation only. Their purpose is to create an atmosphere conducive to innovation and development. Refereeing is less rigorous and papers are subjected to extensive group discussion.

Publications arising from IFIP events vary. The papers presented at the IFIP World Computer Congress and at open conferences are published as conference proceedings, while the results of the working conferences are often published as collections of selected and edited papers.

Any national society whose primary activity is in information may apply to become a full member of IFIP, although full membership is restricted to one society per country. Full members are entitled to vote at the annual General Assembly, National societies preferring a less committed involvement may apply for associate or corresponding membership. Associate members enjoy the same benefits as full members, but without voting rights. Corresponding members are not represented in IFIP bodies. Affiliated membership is open to non-national societies, and individual and honorary membership schemes are also offered.

Electronic Design Automation Frameworks

Volume 4

Proceedings of the fourth International IFIP WG 10.5 working conference on electronic design automation frameworks

Edited by

Franz J. Rammig
Universität-GH Paderborn
Paderborn, Germany

and

Flávio R. Wagner
Universidade Federal do Rio Grande do Sul
Porto Alegre, Brazil

Published by Chapman & Hall on behalf of the
International Federation for Information Processing (IFIP)

SPRINGER-SCIENCE+BUSINESS MEDIA, B.V.

First edition 1995

Originally published by Chapman & Hall in 1995

ISBN 978-1-4757-5106-2 ISBN 978-0-387-34880-3 (eBook)
DOI 10.1007/978-0-387-34880-3

A catalogue record for this book is available from the British Library

Printed on permanent acid-free text paper, manufactured in accordance with ANSI/NISO Z39.48-1992 and ANSI/NISO Z39.48-1984 (Permanence of Paper).

CONTENTS

Welcome to EDAF'94 vii

EDAF'94 Organization viii

PART ONE Transaction-Based Approaches

1 Transaction-based design data processing in the PRIMA framework
T. Härder and N. Ritter 3

2 A configurable cooperative transaction model for design frameworks
A. Meckenstock, D. Zimmer and R. Unland 13

PART TWO Data Management and Tool Integration

3 Configuration management in the STAR framework
H.G. Ribeiro, F.R. Wagner and L.G. Golendziner 25

4 Autocap: an automatic tool encapsulator
J. Câmara and H. Sarmento 35

5 Schema evolution in the STAR framework
M.R. Fornari, L.G. Golendziner and F.R. Wagner 45

PART THREE Data Types

6 Integration of a domain conceptual model with the EDIF information model
R.Y.W. Lau and H.J. Kahn 57

7 Generating abstract datatypes with remote access capabilities
M. Schuetze, B. Schuermann and J. Altmeyer 66

8 Extending view technology for complex integration tasks
M.C. Jones and E.A. Rundensteiner 76

PART FOUR Federated Systems

9 Framework and multi-formalism: the ASAR project
P. Asar 89

10 Enhancing the data openness of frameworks by database federation services
R. Böttger, Y. Engel, G. Kachel, S. Kolmschlag, D. Nolte and E. Radeke 99

11 Interoperability support by integration of EXPRESS models
G. Scholz and W. Wilkes 109

PART FIVE Communication and Distribution

12 Communication in a distributed environment
I. Videira, P. Veríssimo and H. Sarmento 121

13 Basic requirements for an efficient inter-framework-communication
A. Kunzmann and R. Seepold 131
14 Resource-oriented load distribution in a framework environment
J. Schubert and A. Kunzmann 139

PART SIX Design Management

15 Design Flow Management: more than convenient tool invocation
O. ten Bosch, P. van der Wolf and A. van der Hoeven 149
16 Planning and managing multi-disciplinary and concurrent design processes
M.F. Jacome and S.W. Director 159
17 A user model supporting communication in high-level design systems
A. Rockenberg, S. Heymann and U. Westerholz 169

PART SEVEN Special Design Environments

18 A framework for electrical test engineering automation
G. Wirth, R. Wartchow, J. Petterson and S. Bampi 179
19 A compound information model for high-level synthesis
P. Conradi and N. Dutt 189
20 An object-oriented CAE framework for asynchronous circuit synthesis
U. Baake and S. Huss 199

PART EIGHT Standards

21 Design representation in EDIF Version 3 0 0 and CFI Version 1.0
H.J. Kahn 211
22 STEP technology for ECAD databases
F. Buijs and W. Käfer 221

PART NINE Design Representation and Interfaces

23 Towards design formalization to support reuse in ECAD frameworks
J. Altmeyer, B. Schuermann and M. Schuetze 233
24 A generator for graph-based design representations
A. Bredenfield 243
25 The visual interface of the STAR framework
R.S. Mello, L.G. Golendziner and F.R. Wagner 253

Glossary 263

Index of contributors 281

WELCOME TO EDAF'94

In 1987 when the IFIP Working Group 10.2 ("System Description and Design Tools")[1] organized a workshop on "Tool Integration and Design Environments" this was an early experiment. At this workshop, held in Paderborn, Germany, we wanted to give those pioneers a scientific forum, which started to think about open and integrated design environments for electronic design automation.

This event became a success and at the same time framework technology evolved at various locations worldwide. The impact of framework technology on the entire EDA field was further pushed by the creation of the "Common Framework Initiative" (CFI), a cooperation of more than 50 companies and institutes working in this field worldwide. At the same time standards like EDIF and VHDL became an important fact in the industrial practice.

Therefore, IFIP WG 10.2 decided to install a workshop series based on the event in 1987. This biannual event, called EDAF now (Electronic Design Automation Frameworks) is intended to circulate between the most active places in this scientific area. In 1990 Charlottesville, Virginia, USA, has been selected, with standards being a dominating topic discussed at this event. Two years later EDAF came back to Europe, to Bad Lippspringe, Germany. At this time the JESSI Common Frame (JCF) reached its first step of completion. This resulted in a vendor independent, open framework.

In the meantime it has been observed that framework technology is not bound to electronic design only. Nearly the same technology can be used for mechanical engineering or software engineering, just to mention some examples. At the same time Concurrent Engineering became an important topic in industrial reality. It was not too surprising that advanced framework technology turned out to be well suited to support Concurrent Engineering as well.

With all these trends in mind IFIP WG 10.2 was looking for an adequate location for EDAF'94. Porto Alegre, Brazil, turned out to be an excellent choice for this purpose. The beautiful environment of the Brazilian south should encourage fruitful discussions. Most important for the selection of the location however was the continuous scientific work in the area of frameworks and design environments at the Universidade Federal do Rio Grande do Sul in Porto Alegre.

This year database technology became a central topic of the conference. Various approaches including for instance OO-techniques, transaction-schemes, and data representation have been discussed. A topic that emerged during the last years is the federation of frameworks and their cooperation. These interesting problems have been addressed at EDAF'94 as well. "Classical" EDAF topics like design management, case studies, and impact of standards played obviously an important role at this event. To make sure that we are addressing not only paperwork but also real software, a special session has been reserved for demonstrations and vendor presentations. We think that we were able to compile an interesting program from the very good contributions which we obtained from both sides of the Atlantic Ocean.

These proceedings also contain the first draft of a glossary of technical terms in the field of electronic design automation frameworks, combining contributions from the participants of the event and from the JESSI Common Framework Project.

Franz J. Rammig
Program Chairman

Flávio Wagner
General Chairman

[1]In September 1994, the Working Group 10.2 merged with the former Working Group 10.5 ("VLSI") into a new Working Group 10.5, called "Design and Engineering of Electronic Systems".

EDAF'94 Organization

General Chairman	Flávio R. Wagner, Brazil
Program Chairman	Franz J. Rammig, Germany
Secretariat	Lourdes Tassinari, Brazil

Co-sponsoring institutions

- CFI - CAD Framework Initiative
- ECSI - European CAD Standardization Initiative
- GI - Gesellschaft für Informatik
- IEEE Computer Society
- ITG - Informationstechnische Gesellschaft
- SBC - Brazilian Computer Society

Program Committee

Elfriede Abel	GMD, Germany
Sigfried Buck	Siemens Nixdorf, Germany
Don Cottrell	CAD Framework Initiative, USA
Steve Director	Carnegie-Mellon Univ., USA
Nikil Dutt	Univ. of California at Irvine, USA
Jim Heaton	ICL, UK
Jan Janse	Philips, The Netherlands
Mike Newman	Commission of the European Union, UK
Franz J. Rammig	Univ.-GH Paderborn, Germany
Elke Rundensteiner	Univ. of Michigan, USA
Anton Sauer	European CAD Standardization Initiative, Germany
Bernd Steinmüller	Cadlab, Germany
Jean-Pierre Tual	Bull, France
Rene van Leuken	Delft Techn. Univ., The Netherlands
Massimo Vanzi	ACCENT Sal, Italy
Flávio R. Wagner	Univ. Rio Grande do Sul, Brazil
Ron Waxman	Univ. of Virginia, USA

PART ONE

Transaction-Based Approaches

1

Transaction-Based Design Data Processing in the PRIMA Framework

Theo Härder, Norbert Ritter
Department of Computer Science
University of Kaiserslautern
P.O.Box 3049, 67653 Kaiserslautern, Germany
e-mail: haerder|ritter@informatik.uni-kl.de

Abstract

Nowadays database technology is a key concept of integrated design environments and CAD frameworks. Crucial aspects of the exploitation of database systems in this application area are the modeling of design data as well as the modeling of activities managing this data. The latter point requires the development of transactional structures reflecting the actual processing characteristics of the various classes of design data in an adequate manner and, additionally, some support for controlled cooperation among designers. In this paper, we will discuss the distinct processing characteristics of the most important types of design data and will propose adjusted transactional structures reflecting the requirements of their manipulation. Furthermore, we will detail how these transactional structures can be embedded into a comprehensive processing model for design processes, which also supports cooperation control and design flow management besides data processing performed by a set of design tools.

1 Introduction

The term *CAD framework* [HNSB90, RS92] comprises all facilities supporting the CAD tool developer, the CAD system integrator as well as the system user (designer). Hence, the aspects of design data representation and design dynamics support are crucial issues. Design data representation calls for an adequate management of multiple types of data which can be classified by structural criteria (flat tuples vs. complex structured data) as well as application-specific criteria (versioned data vs. version-free data) [Ka90, WGLL92]. Capturing design dynamics requires a powerful processing model, which on one hand supports cooperation between designers and on the other hand keeps the design consistent. To accomplish the various aspects of data consistency, the approved transaction processing concept of database systems (Atomicity, Consistency, Isolation, Durability) [HR83] can be applied; it guarantees atomicity and serializability for a transaction which, for this purpose, isolates its execution from other transactions. For this reason, cooperation - a fundamental prerequisite in CAD frameworks - is not a natural property of ACID transactions. However, it may be supported by introducing a hierarchically structured processing model where ACID transactions are essential building blocks of the control structure. To solve cooperation control and data consistency, we propose to enhance the approved transaction concept by explicit cooperation control facilities and by further mechanisms supporting design management and design flow management.

PRIMA Framework

PRIMA Framework is a CAD framework which uses the structurally object-oriented database system PRIMA [Sch93] as integrated design data repository. According to [HNSB90] a CAD framework offers a number of services which are supported by the PRIMA framework as subsequently outlined[1]. On top of PRIMA, we implemented an object and version management system [KS92] providing means for the manipulation of explicit complex-object versions as well as for the manipulation of configurations (*version and configuration services*). The object and version management system offers the Object Query

[1] We will not consider the most elementary framework services of process management and physical data management constituting a common interface between the operating system and the framework itself.

Language (OQL) at its interface which is embedded into a higher programming language. An adequate processing concept providing means for workstation/server communication and processing of cached design data on the workstation side facilitates the direct implementation of design tools as well as the implementation of foreign tool interfaces (***tool integration services***). During the design process, the design tools can be applied in a controlled manner. To pre-plan and schedule design tool applications ***design-flow management services*** are provided. The intention of these services is to guide users through complex activities and to apply certain design methodologies. Design-flow specifications can be (re-)used in multiple design tasks or design processes (***design methodology services***). Further services, e.g. ***user-interface services***, are currently elaborated, but are beyond the scope of this paper.

In this paper, we focus on information sharing in cooperative work arrangements and the corresponding activity control aspect. Thus, we will first outline the major concepts of the object and version data model (Sect. 2). Afterwards an overview of the CONCORD processing model will be given (Sect. 3) capturing design dynamics and, therefore, comprising data management services (transactional concepts), design-flow management services as well as design-management and cooperation-control facilities. In Sect. 4, we will discuss transactional control structures allowing for the processing of design data. Regarding the processing characteristics of the data we will select the most adequate transactional structure and we will describe how this structure is embedded into the comprehensive CONCORD model. The last section (Sect. 5) gives a conclusion.

2 Object and Version Data Model (OVDM)

In [KS92], the object and version data model (OVDM) is proposed for the management of design data. We will only be able to outline the major concepts of this model, illustrated in Fig. 1.

Complex Objects and Versions

Complex Objects are identifiable occurrences of complex-object types and are given as structured sets of elementary data. In its entirety, a complex object is described by object attributes. An (complex) object combines ***elementary objects***, which can be compared with the tuples of the relational model. In addition to the features of the relational model, elementary objects can be connected via (typed) ***structural relationships***. Obviously, it is one of the major tasks of a design process to create "contents" of objects, i.e., nets of elementary objects. Design is an iterative process, which typically leads to several (similar) nets of elementary objects, which we call ***versions***. Thus, versions are different states of the net of elementary objects constituting a complex object. In this way, versions are capturing the various object states derived during the design process with the intention of a stepwise improvement of preliminary data in order to reach the (partial) design goal. The relationships between the versions of a single object, representing the derivation of new versions out of existing versions, are managed in form of a derivation graph, which, in turn, can be organized as list, tree, or acyclic graph.

Relationships between Objects / Versions

Objects can be connected by ***object relationships***. Obviously, these relationships must also be captured at the version level. Therefore, ***version relationships*** can be seen as refinements of object relationships. Version relationships depend on the existence of relationships between the corresponding objects. In addition to the explicitly represented version relationships implicit relationships can be modeled resulting from data overlapping. Overlapping can occur directly by shared elementary objects or by inter-structural relationships which should only be interpreted in configurations for consistency reasons. In Fig. 1, we see that version 3 of object 1 is connected with versions 1 and 3 of object 2 by version relationships. Remember that every version is representing its object completely. This implies that version 3 of object 1 actually can only be in relationship with either version 1 or version 3 of object 2. Such integrity constraints are checked at the level of configurations.

Configurations

It is the goal of configuring to establish consistent units by selecting certain versions out of the set of versions stored in the database [Ka90]. ***Configurations*** (not shown in Fig. 1) are occurrences of spec-

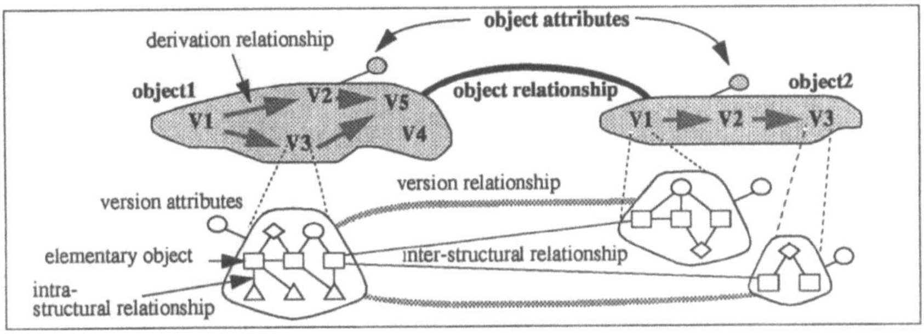

Figure 1: Major Concepts of the OVDM

ified configuration types which are usually associated with special integrity constraints expressing the requirements of the unit of versions to be established during the configuring activity. Configuration types are defined on the basis of object types and relationship types. A configuration type is given as a tree where the nodes represent complex-object types and edges embody paths which, in turn, may consist of complex-object types and relationship types. The configuring process starts with a version of the root type. For every inner node, a version selection must be done satisfying certain requirements. For example, it can be enforced that at a certain node all selected versions must belong to distinct objects or that certain semantic integrity constraints must be fulfilled on the selected versions or the resulting structure, respectively. Furthermore, an important property of configuration types is their structure, i.e., hierarchical, or even recursive.

The OVDM concepts, outlined in this section, are used in the PRIMA framework to model design data. In the following section, we will give an overview of the activity model used to process these data structures.

3 CONCORD Processing Model

The CONCORD[2] model [Ri94a] captures the dynamics inherent to design processes. To reflect the spectrum of requirements, such as *hierarchical refinement, goal orientation, stepwise improvement and team orientation*, three different levels of abstraction are distinguished as roughly illustrated in Fig. 2.

Administration/Cooperation Level (AC Level)

At the highest level of abstraction, we reflect the more creative and administrative part of design work. There, the focus is on the description and delegation of design tasks as well as on a controlled cooperation among the design tasks. The key concept at this level is the ***design activity (DA)***. A DA is the operational unit representing a particular design task or subtask. During the design process, a ***DA hierarchy*** can be dynamically constructed resembling (a hierarchy of) concurrently active tasks. All relationships between DAs essential for cooperation are explicitly modelled, thus capturing task-splitting (cooperation relationship: ***delegation***), exchange of design data (cooperation relationship: ***usage***), and negotiation of design goals (cooperation relationship: ***negotiation***). The inherent integrity constraints and semantics of these cooperation relationships are enforced by a central system component, called ***cooperation manager***.

Design-Control Level (DC Level)

Looking inside a DA reveals the DC level. There, the organization of the particular actions to be performed in order to fulfill a certain (partial) design task is the subject of consideration (***design flow***). At this level, Fig. 2 shows an execution plan (***script***) of a particular design activity. The corresponding

[2]The CONCORD acronym stands for: CONtrolling COopeRation in Design environments.

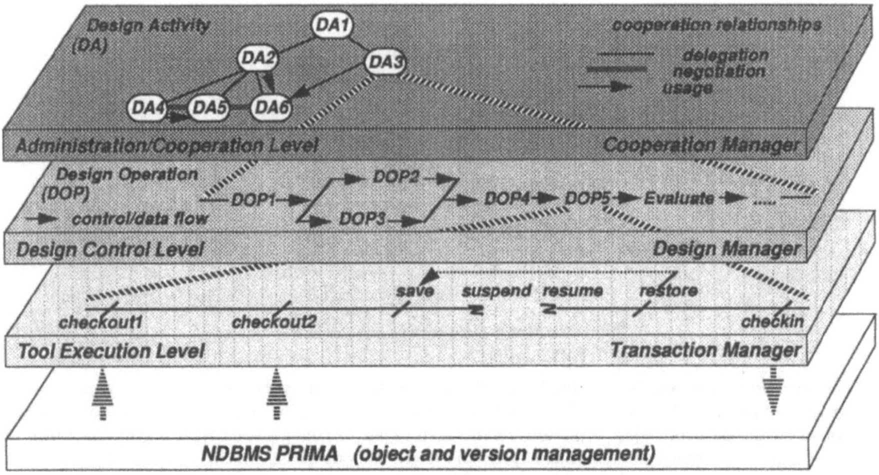

Figure 2: Abstraction Levels of the CONCORD Model

script models the *control/data* flow among several design actions performed within a DA. Usually, these actions are design tool applications. The operational unit serving for the execution of a design tool is the *design operation (DOP)*. In order to control the actions within the scope of a single DA, but without restricting the designers' creativity, flexible mechanisms for specifying the design flow for a DA (scripts, constraints, event-condition-action rules) are provided. The correctness of tool executions is guaranteed by a system component, called *design manager*. The design manager does also provide for recoverable script executions that is needed for level-specific and isolated failure handling. Design tools are applied to improve existing design states in order to finally reach at a design state that completes the current (partial) design task. Design states are captured by means of the object and version model introduced in the previous section. The derivation of versions and corresponding design object states by means of tool applications is supported by the concepts provided at the TE level.

Tool-Execution Level (TE Level)

From the viewpoint of the DBMS or data repository, a DOP is a long transaction (see in Fig. 2). A DOP has the properties of conventional transactions. Because of long duration, it is internally structured by save/restore and suspend/resume facilities [HHMM88] to be able to rollback the design at the application level and to continue the design work after breaks. A DOP processes design object versions in three steps. First, the input versions are checked out from the integrated data repository and cached in an object buffer at the workstation for efficiency reasons. Second, the design data is mapped to storage structures tailored to the application needs. It is processed by one or more design tools. Third, the finally derived new versions are propagated back to the data repository (check-in operation). The derivation of schema-consistent and persistent design object versions is guaranteed, again, by a central system component, called *transaction manager*. It is also responsible for the isolated execution of DOPs and for recoverable DOP executions that are, again, necessary for a level-specific and isolated failure handling. The transaction manager employs mechanisms provided by the advanced DBMS which manages the integrated data repository. Since long transactions are involved, incompatible lock requests on data already locked for an anticipated long duration[3] cannot be handled by usual wait mechanisms, but require a notification concept.

[3]In the case of version derivations, generally long locks are not required, because multiple successors of a version can be derived concurrently.

Overall Architecture

At the conceptual level, all information relevant for a design process is managed by one logical server so that no exchange of design data between distributed databases becomes necessary during the design process and a cooperative data exchange between concurrently active design tasks can be managed via the integrated data repository. The logical server, in turn, may be implemented by a set of distributed and heterogeneous physical servers; this mapping has to be achieved by the DBMS which encapsulates the interoperability problems from the logical server view of design data management. In contrast, the designer carries out his design work at a (single- or multi-processor) workstation. Since a DA typically comprises the design work of a single designer, we assume that a DA is running at a single workstation. Consequently, all actions executed within a DA are managed and executed at this workstation, too. This is needed for three reasons. Firstly, in many cases the designer has to specify input parameters for the design tools. Secondly, designer interaction during tool execution is necessary and essential, and, thirdly, the information derived by a DOP is mostly subject to work-flow (data-flow) management within the DA.

So far, we have outlined data modeling and activity modeling aspects as well as some architectural issues. In the following section, we will examine the needs of the application pertaining the modeling and the management of various classes of design data in more detail. This will lead to a refinement of the transactional control structures used in the CONCORD model.

4 Transaction-Based Design Data Processing

In this section, we want to discuss transaction-based processing of design data in a little more detail. CONCORD provides a very coarse view to design data processing. The processing model must be refined regarding the various types of design data accessed during the design process. Therefore, we first want to classify the most important types of design data. Afterwards, we will detail the CONCORD processing model.

First of all, however, we have to redefine the notions of transaction commitment and data release, because they are used in slightly different meaning. Committing a DOP means making schema-consistent data persistent, but it does not imply that this data becomes automatically visible to any user. It remains associated to the corresponding DA. Exchanging data between DAs or making data visible to other DAs is captured by the notion of data release. Here, two cases can be distinguished. First, with the successful termination of a DA, its results (final design data of the DA) are passed to the parent DA. Second, preliminary design data can be exchanged between DAs via the usage relationship enforcing a special cooperation protocol. Final design data of the overall design process becomes generally visible with the successful termination of the top-level DA.

Classes of Design Data

During a design process two major classes of design data are accessed. We call the first class *task-related data (DA-related data)*. This data can be characterized as follows:

- it is subject of design work (stepwise improvement); it represents the design objects of DAs and captures the different design object states developed during the design;
- it is usually complex structured, versioned and manipulated by long transactions exploiting check-out/ check-in primitives; it is useful to manage different states of task-related data in order to be able to return to an older design state as a starting point for the development of a better solution;
- it is subject to various kinds of cooperation; first, there is a restricted release of the final design data of a DA to the parent DA (along the delegation relationship); second, preliminary design data can be exchanged among DAs (via usage relationships); third, design objects can be shared among certain DAs (*object pools*); they can manipulate them in a mutual manner;
- the design process is terminated by an unrestricted release (successful termination of the top-level DA); if the data has to be further protected, access control mechanisms seem to be appropriate.

The second class (***common design data***) comprises the data which is not related to a single design process. For example, information libraries, general rules, regulations, and catalogs about standards belong to this class. Furthermore, information about technology, tolerances, materials, etc., is contained in this class. This data is typically accessed in the following way[4]:

- read access is predominating; modifications are infrequent;
- information requests can usually be handled by means of short transactions;
- there can be concurrent and independent requests of a single designer in parallel to his design tool applications.

However, upon arrival of new rules, standards, etc., or when errors are detected, the common design data is occasionally updated by the database administrator or even by the design transactions. These changes have to be performed atomically.

Before detailing the CONCORD processing model in order to be able to adequately process both classes of design data, we want to outline how CONCORD exploits OVDM concepts to manage task-related data.

Interplay of CONCORD and OVDM

The discussion of the characteristics of the task-related data indicates that CONCORD and OVDM are closely interplaying. The design objects manipulated by the DAs of a DA hierarchy (or their DOPs, respectively) can adequately be modelled by means of OVDM concepts.

At creation time, every DA is associated with a ***design object type***. The design object type is usually given as a set of complex-object types which are connected by object relationship types forming a configuration type. Thus, the objects and versions manipulated by a certain DA in order to reach its design goal are occurrences of the complex-object types contained in its design object type. This data is processed within DOPs. A DOP receives as input a set of versions, does some computations, and produces as output a schema-consistent set of versions (see TE level). We are considering this set of versions as an implicit (and mostly partial) ***design object state*** of the corresponding DA. An explicit (and complete) design object state is given as an occurrence of the configuration type constituting the design object type of the DA[5]. The creation of preliminary and final design object states as well as the synthesis of partial results delivered by sub-DAs can adequately be done by means of the mentioned configuration concept. Usually, the hierarchical structure of configuration types (sub-configurations) is a natural basis for the decomposition of design tasks associated with DAs.

Detailing the Processing Model

In distributed environments (such as workstation/server architectures), there exist two fundamental principles to process data: either ***the functions follow the data*** or ***the data has to be moved to the functions***. The processing characteristics of the data determines, which principle has to be selected to achieve optimal performance. For this reason, adjusted processing models have to observe the predominating properties of data access.

Considering the processing structures in a little more detail, we have to incorporate control structures for accessing task-related data as well as common design data. From the viewpoint of the DAs, common design data is typically subject to information requests. Despite this predominating read access we also have to consider updates on common design data initiated by DAs. In every case, however, it is not necessary to cache and process common design data at the workstation, because locality of references cannot be anticipated in contrast to the processing of task-related data. Thus, we have the following aspect distinguishing the processing of the two mentioned classes of design data. The manipulation of

[4] Archived design data (from earlier design processes) has similar processing characteristics.

[5] Since the creation of configurations is expected to be done in the later phases of the design process, we are also considering the results of DOPs as preliminary design object states which in turn can be subject of cooperation between DAs.

task-related data can adequately be supported by caching it at the workstation and providing an appropriate processing concept for navigational and repetitive access (e.g., hierarchical cursors [HS92]). The manipulation of common design data, in contrast, is best performed by sending a function call to the server, which then computes the function using the stored data and sends back the results to the corresponding workstation.

So we have to raise the question on how to adequately organize a DA's access to the various types of design data. Regarding the needs of the application, we first of all have to consider, whether it is sufficient to provide special and independent application programs allowing the designer to do information requests (separate processes for information requests and DOPs[6]) or whether it is necessary to incorporate these information requests into the tools processing the task-related design data (a single process). In any case, appropriate transactional structures have to be supported [SI93].

In the following, we will examine different transactional structures in order to find an adequate one for the *combination of DOP processing and information requests.*

A single long transaction. Handling information requests within the long transaction processing the taskrelated data leads to a very simple transactional structure. The drawbacks are obvious. The locks on the common design data must be held until commit of the long transaction in order to provide correct (serializable) schedules [BSW79]. This could be an unnecessary long period during which blocking of concurrent (long) transactions is not acceptable and which would require frequent notifications even for accesses to common design data.

A single nested transaction. The nested transaction model [Mo81] allows the creation of a transaction tree. The subtransactions are atomic and are isolated against each other. Thus, the model allows for the parallel execution of the actions performed within sibling transactions and for a fine granular recovery since subtransactions can be aborted without any effect on the parent or sibling transactions. Only the top-level transaction has all ACID properties, because all locks hold by a committing subtransaction are upward inherited. This model would allow to perform actions processing task-related data and actions on common design data in separate subtransactions so that the abort of one transaction would not affect the other actions. Despite this higher flexibility we still have the drawback that all locks, and in particular the locks on common design data, are to be held till commit of the top-level transaction.

A single nested transaction with open subtransactions. An extension of the nested transaction model (inspired by the sagas model [GS87]) proposes to distinguish closed subtransactions from open subtransactions. Closed subtransactions are those of the original model passing the locks to the parent transaction at commit. Open subtransactions, on the other side, release their locks at commit (of the subtransaction) so that the corresponding data becomes accessible for concurrent (nested) transactions. This model allows for handling access to common design data within open subtransactions so that the corresponding common design data can be released at commit of the subtransaction. The drawback of this model is that it requires a compensation transaction for every open subtransaction allowing for a semantic UNDO of the subtransaction after its commit. This becomes necessary, whenever a higher level transaction aborts. This abort dependency would be only adequate, if there would be causal dependencies from the information request to the manipulation of task-related design data. We don't see these causal dependencies. For that reason, it is not necessary to provide compensation transactions so that this model is not adequate, too.

A single activity/transaction hierarchy. The ATM (activity/transaction model [DHL91]) provides very flexible mechanisms for the creation of activity/transaction hierarchies. This model enhances the mentioned nested transaction model. Additional operational units called ***activities*** can be incorporated which do not have transactional properties. For example, accesses to read-only data (bulletin board, etc.) could be performed in activities as well as operations not related to database data. Activities and transactions can be arbitrarily nested except nesting activities into transactions. In this way, pure nested transactions are establishing subhierarchies of an activity/transaction hierarchy. The relationship

[6]We use the term "DOP" synonym to "manipulation of task-related data" and "information request" synonym to "manipulation of common design data".

resulting from nesting a subtransaction into an activity is characterized as follows: The transaction has access to all objects manipulated by the activity and the activity can only terminate after the termination of the transaction. A further feature of this model is the possibility to define causal dependencies between activities/transactions. If unit A is causally dependent from unit B this implies that ABORT(B) will lead to ABORT(A). There are two ways to model the manipulation of design data by the concepts of the ATM model. The first way is to treat both, the manipulation of task-related data as well as the manipulation of common design data, as (sub)transactions. As already mentioned, there are no causal dependencies from the manipulation of common design data to the manipulation of task-related data, so that this feature cannot be exploited and the model would degenerate to the pure nested transaction model which is not adequate for the reasons mentioned above. The second possibility is to treat access to common design data as activities and not as transactions. This is also not appropriate because the manipulation of common design data should also be protected by transactional mechanisms.

Multiple independent transactions. The last transactional structure for combining the manipulation of taskrelated data and the manipulation of common design data is a very easy one providing high flexibility. While the access to common data is performed within a short ACID transaction the processing of task-related data is encapsulated by an independent long transaction. This simple structuring ensures that the different processing characteristics do not influence each other negatively. For that reason, we have chosen this last possibility. It enforces to mark the transaction types already during the implementation of the application programs and design tools because the distinct types of transactions are implying distinct processing models. While the long transaction, which is well suited for the processing of task-related data, usually complex structured and versioned, exploits check-out/check-in primitives as well as caching and processing of design data at the workstation side (see properties of DOPs), the conventional short transaction, which is well suited for the manipulation of common design data, can be seen as a function call which is executed at the server and causes only the results to be transferred back to the workstation. In order to support adequate supply of design tool applications (encapsulated by long transactions) with context information as well as information derived from common design data, adequate parametrization of design tools and adequate data-flow capabilities must be provided.

Data Flow

The intuitive model of data flow in CONCORD is that long transactions propagate the newly derived versions to the data repository by check-in operations so that data supply of succeeding transactions can be managed via check-out operations. This model does not provide good performance in the case where the data produced by a certain tool is needed as input for a directly succeeding tool. Therefore, we propose the following optimization. The new versions produced by a DOP, which are needed by a succeeding DOP, remain in the workstation cache, so that they can be further processed by the succeeding DOP. In order to provide durability of these versions we have to write a persistent log at the workstation side. The server just gets the information that new versions are derived and generates the according version numbers and housekeeping information. By this mechanism, it is possible to asynchronously transfer the new versions to the server (e.g., at the latest, when they are accessed by a cooperating DA performed at another workstation). The mentioned actions for check-in of a DOP are appropriate, because the results of long transactions remain task-related, although they are to be stored persistently in the data repository. The results remain associated to the corresponding DA and can only be made visible to other DAs via cooperation primitives.

Concurrency/Cooperation-Control

As mentioned previously all (short and long) transactions, independently to which DA they belong, are executed in isolation. Short transactions are manipulating common design data at the server side and their results are released at commit. Long transactions (DOPs), on the other hand, are processing their data at the workstation side, their results remain task-related and the data release depends on the cooperation relationships of the corresponding DA.

Regarding versioning, synchronizing long transactions becomes quite easy. As compared to the usual readwrite synchronization on version-free data, versions allow for much higher concurrency as illustrated by the lock compatibility matrix in Fig. 3. Generally, the concurrent derivation of several successors of the same input version does not lead to any conflicts. Despite of that, it may be necessary in special

granted mode \ requested mode	S	D	XD	X
S	+	+	+	-
D	+	+	-	-
XD	+	-	-	-
X	-	-	-	-

lock modes:
S: needed to read a version;
D: needed to derive a successor;
XD: exclusive successor derivation;
X: needed to update a version;

Figure 3: Lock Compatibility Matrix for Version Manipulations

cases to prevent parallel derivations from the same input version for application-specific reasons. Here, as well as in the case of version updates (which occur very seldom) blocking is required. In order to avoid an unreasonably long waiting situation, which can be anticipated whenever a check-out operation cannot acquire the required read or derivation locks, this situation is reported to the tool. This allows the tool implementor to foresee appropriate contingency actions, e.g., notification of the designer.

As already mentioned at the beginning of this section, the release of task-related data depends on the cooperation relationships of the corresponding DA hierarchy. Especially along usage relationships the creating DA may permit cooperating DAs to access propagated design data. The kind of access to co-operatively exchanged data depends on the chosen protocol. For example, it could be sufficient to only permit read access; in other cases, it could be appropriate to let the cooperating DA derive new versions and to give them back as some kind of proposal. Since the discussion of these protocols is beyond the scope of this paper (see [Ri94a, Ri94b]), we just want to mention that the server has to manage the cooperation relationships between the DAs. Thus, the server is aware of the release state of task-related data and can decide whether a certain DA's DOP can be permitted to access a requested version.

5 Conclusions

In this paper, we have been focusing on the data modeling and data manipulation aspects in cooperative design applications. The CONCORD processing model which allows the straightforward mapping of the processing structures predominating in design processes and, therefore, serves as the processing model of the PRIMA framework, has been refined so that processing of the most important types of design data can adequately be encapsulated within ACID transactions. The concepts of the object and version data model OVDM are well suited to model these classes of design data. Thus, CONCORD and OVDM are closely interplaying in order to meet the requirements of the application area.

Furthermore, CONCORD incorporates cooperation control and design-flow capabilities. To support collaboration explicit control mechanisms supporting a user-centered cooperation [Ri94b] are combined with traditional concurrency control mechanisms. This, on one hand, allows a very flexible cooperation among designers, and, on the other hand, keeps the design data consistent, because approved transaction concepts are applied. The design-flow features could not be detailed in this paper. This matter concerns pre-planning and scheduling complex activities in order to guide the user and help him to concentrate on the creative design decisions he has to take. We are currently working on this topic in order to find appropriate control structures to connect the transactional steps processing design data (cf. [WR92], [RW92]).

The aspects mentioned in this paper address major objectives of the DoCAIA project (Brazilian/German Cooperation on the "Support of Design and Development of Computer Applications in Industrial Automation"), so that the results discussed in this work can be a foundation of a further cooperation in the context of this project.

References

[BSW79] Bernstein, P.A.; Shipman, D.W.; Wong, W.S.: Formal Aspects of Serializability in Database Concurrency Control, IEEE Transactions on Software Engineering, Vol 5, No. 5, May 1979, pp. 203-216.

[DHL91] Dayal, U., Hsu, M., Ladin, R.: A Transactional Model for Long-Running Transactions, Proc. 17th Int. Conf. on VLDB, Barcelona, Spain, 1991, pp. 113-122.

[GS87] Garcia-Molina, H.; Salem, K: Sagas, Proc. ACM SIGMOD, 1987, pp. 249-259.

[HHMM88] Härder, T., Hübel, C., Meyer-Wegener, K., Mitschang, B.: Processing and Transaction Concepts for Cooperation of Engineering Workstations and a Database Server, Data and Knowledge Engineering 3, 1988, pp. 87-107.

[HNSB90] Harrison, D., Newton, R., Spickelmier, R., Barnes, T.: Electronic CAD Frameworks, Proc. of the IEEE, Vol. 78, No. 2, Febr. 1990, pp. 393-417.

[HR83] Härder, T., Reuter, A.: Principles of Transaction-Oriented Database Recovery, ACM Computing Surveys 15, 4, 1983, pp. 287-318.

[HS92] Hübel, C., Sutter, B.: Supporting Engineering Applications by New Data Base Processing Concepts - An Experience Report, Engineering with Computers, 8, pp. 31-49, 1992.

[Ka90] Katz, R.: Toward a Unified Framework for Version Modeling in Engineering Databases, ACM Computing Surveys, Vol. 22, No. 4, Dezember 1990, S. 375-408.

[KS92] Käfer, W., Schöning, H.: Mapping a Version Model to a Complex Object Data Model, Proc. 8th Int. Conf. on Data Engineering, Tempe, Arizona, 1992.

[Mo81] Moss, J.E.B.: Nested Transactions: An Approach To Reliable Computing, M.I.T. Report MIT-LCS-TR260, M.I.T., Laboratory of Computer Science, 1981.

[Ri94a] Ritter, N., Mitschang, B., Härder, T., Gesmann, M., Schöning, H.: Capturing Design Dynamics - The CONCORD Approach, Proc. 10th Int. Conf. on Data Engineering, Houston, Texas, Feb. 1994, pp. 440-451.

[Ri94b] Ritter, N.: An Infrastructure for Cooperative Applications based on Conventional Database Transactions, Proc. of the CSCW Infrastructure Workshop, Chapel Hill, North Carolina, Oct. 1994.

[RS92] Rammig, F. J., Steinmüller, B.: Frameworks and Design Environments, in: Informatik-Spektrum (1992) 15: 33-43 (in German).

[RW92] Reinwald, B., Wedekind, H.: Automation of Control and Data Flow in Distributed Application Systems, in: Tjoa, A., Ramos, I. (eds): Database and Expert Systems Applications - DEXA (Proc. of the Int. Conference, Valencia), Springer, Wien, 1992, pp. 475-481.

[Sch93] Schöning, H.: Query-Processing in Complex-Object Database Systems, Deutscher Universitäts-Verlag, 1993 (in German).

[SI93] Souto, M.A.M., Iochpe, C.: Transaction-Based Support for Production Plan Execution: a Position Paper, Proc. 1st Brazilian Symposium on Intelligent Automation, Aguas Claras, SP, Brazil, 1993.

[WGLL92] Wagner, F.R., Golendziner, L.G., Lacombe, J., Lima, A.V.: Design Version Management in the STAR Framework, in: Electronic Design Automation Frameworks, Newman, M. and Rhyne, T. (eds.), Elsevier Science Publ. (North-Holland), 1992, pp. 85-97.

[WR92] Wächter, H.; Reuter, A.: The ConTract Model, in: Transaction Models for Advanced Database Applications, Morgan Kaufmann, San Mateo, CA, 1992.

2

A Configurable Cooperative Transaction Model for Design Frameworks

Axel Meckenstock
Detlef Zimmer

CADLAB*
Bahnhofstr. 32
D-33102 Paderborn
{axel|det}@cadlab.de

Rainer Unland

Westfälische Wilhelms-Universität
Grevener Str. 91
D-48159 Münster
unlandr@uni-muenster.de

1 Introduction

Traditional transaction management [Gra81] which supports concurrent work and failure handling (recovery) based on the *ACID*-properties[1] [HR83] plays a significant role in database management systems. In the field of design frameworks this concept is relevant, too. However, the set of requirements for transaction management in this area is much more diverse.

In the conventional model, **concurrency control** is done by executing transactions in an isolated way. Design processes, however, require different degrees of **cooperation** in order to support collaborative work. **Recovery**, which is traditionally handled by rolling back transactions, should consider that design activities are of long duration and that work losses should be minimized. **Consistency** is handled implicitly in the conventional model. In design environments, explicit consistency specifications and different levels of consistency should be supported.

The idea of design frameworks is based on the observation that "hard-wired" services are not flexible enough to meet the various requirements of applications. Instead, a framework should offer a high degree of configurability. This idea can also be applied for transaction management. Thus, we propose a **highly configurable transaction model** permitting the definition of different **transaction types**. Hereby, a heterogeneous transaction hierarchy can be formed [US92, MUZ94]. The transaction types can be assigned different characteristics with respect to concurrency control, cooperation, recovery and consistency management. This is done by defining **protocols** that are used by the corresponding transactions. It is the task of the framework administrator to define suitable transaction types fulfilling the application requirements. Framework users and/or applications needing transactions with certain characteristics can choose from the set of transaction types and need not be bothered with details of transaction management.

---

*Joint R&D Institute University-GH Paderborn / Siemens Nixdorf Informationssysteme AG.

[1] Atomicity, Consistency, Isolation, Durability

We shall start with a short review of related work. In section 3 we present the basic concepts of our transaction model. Section 4 describes the main idea, i.e., the configurability of the model. We conclude with a short summary and an outlook.

2 Related Work

Early approaches to design transaction management have concentrated on certain aspects. For example, [KLMP84] introduced the concept of workspaces and *checkout/checkin*, [KSUW85] added version management and some cooperation primitives, and [KKB88] presented a model consisting of four transaction types and defined concurrency control protocols permitting co-operation. [Ioc89] discusses recovery techniques in workstation-server environments for several design transaction models. [NRZ92] uses grammars as a programmable correctness criterion for cooperative transaction hierarchies.

A more recent approach, *Concord* [RMH+94], especially deals with cooperation. Although it uses similar notions of transactions as our approach, it differs, e.g., in the way the operations *checkout/checkin* are handled. Another model developed within the *JESSI Common Framework* project [BS94] describes primitives for design transactions on top of an object-oriented database system. Both approaches do not support the concept of typing of transactions.

The main benefit of our model lies in the ability to configure transactions. Hereby, heterogeneous transaction hierarchies can be built that satisfy various requirements of applications. This heterogeneity also allows to combine the best-suited concepts from other transaction models. Furthermore, the model is supposed to integrate the different aspects of transaction management, in particular concurrency control, cooperation, recovery and consistency management. In this way, we continue and generalize the *transaction toolkit* approach [US92].

3 The Transaction Model

3.1 Overview

In this section we present the basic concepts of our transaction model. In particular, we sketch three notions of "transactions" which is necessary since this term is overloaded in literature.

To illustrate our presentation we give a simple example that will be used throughout this paper. A chip design project has the task to build an *arithmetic-logical unit* (*ALU*). The task is subdivided into the *design* and the *simulation* of the *ALU*. The design of the *ALU* can be further partitioned into the design of submodules like *adders* and *multipliers*. Designers are supported by interactive or batch tools, e.g., a `schematic editor`, a `netlist generator`, or a `simulator`. These tools store design objects (e.g., schematics or netlists) in a database. They perform operations like reading a schematic into a main memory buffer, writing it back, inserting new modules into a schematic or adding a link into a netlist.

From these observations we can derive three kinds of transactions: **Design Transactions (DT)** are used to model certain design tasks, **Tool Transactions (TT)** represent the execution of tools, and **Atomic (Database) Transactions (AT)** perform the elementary operations on the database.

3.2 The Elements of the Model

We illustrate the model by the schema depicted in fig. 1 and the example in fig. 2.

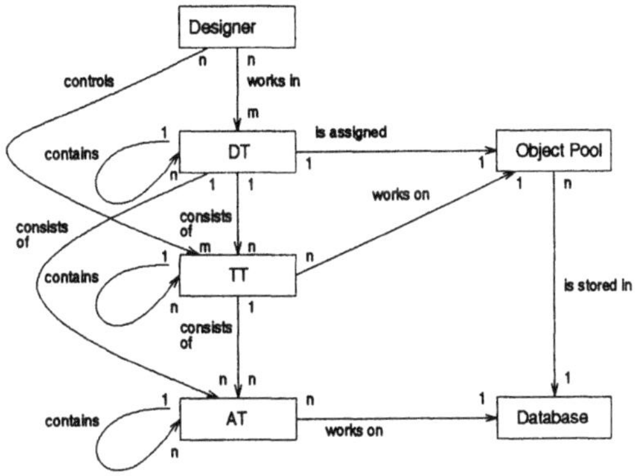

Figure 1: Schema for the Transaction Model

A **Designer** works on design tasks or controls their execution. If he is participating in several tasks, he is assigned several DTs. Vice versa, several designers can be assigned a single DT if they cooperate closely and if the task cannot be split in a reasonable way. Thus we get a n:m-relation between designers and DTs.

A **DT** represents a design task. Often design tasks are subdivided into subtasks, resulting in a hierarchy of DTs. The design objects manipulated by a DT are typically managed as local copies. This leads to a workspace concept that distinguishes between (semi-)public and private workspaces [KLMP84]. To realize such a concept, we introduce **object pools** [US92]. An object pool is assigned to a DT and serves as a (logical) container for all objects accessed by the DT. The operations *checkout* and *checkin* are used to copy objects between object pools. By using these operations DTs can cooperate *explicitly*. DTs are of long duration and typically do not satisfy the *ACID*-properties. Concurrency control is done by a persistent mechanism (e.g., persistent locks) spanning sessions. Recovery must be done in a flexible way since a total rollback of long-lived design activities is often not adequate. In case of a crash, DTs can be reconstructed and continued, because the actual work is done by TTs and ATs, which store their results in a persistent way.

An object pool contains the **design objects** manipulated by a DT. Optionally, objects may be versioned in order to represent the design history or variants. We distinguish between the object pool as the logical workspace and the **database** as the physical container. The fact that data will typically be distributed within a workstation-server environment is not relevant for the discussion in this paper and will therefore be ignored. We assume that objects are manipulated within transaction boundaries and that each elementary operation is performed by an AT which obeys the *ACID*-properties.

Within a DT **tools** like editors or simulators are executed. For simplicity we assume that a **TT** represents the execution of exactly one tool. The object pool of a DT serves as the logical data repository for the TT. The DT has to ensure (in charge of the user or a TT) that needed objects are available in the object pool with appropriate access rights. TTs can be of short or

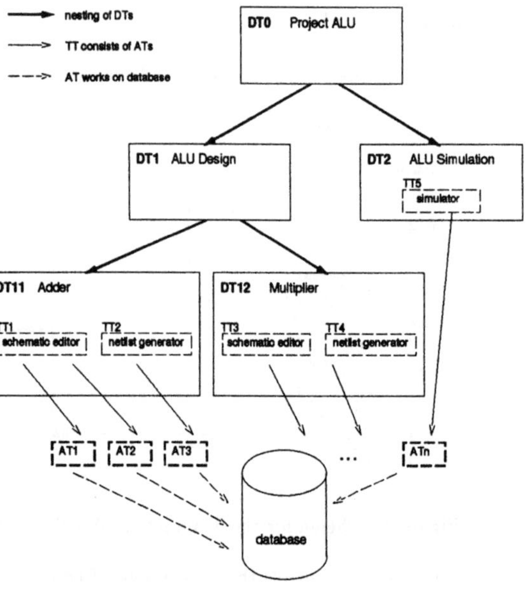

Figure 2: Example

long duration and may ensure the *ACID*-properties. If a TT spans several designer's sessions, it can be suspended and resumed later. TTs are not necessarily units of consistency since a single tool execution need not bring a design object into a consistent state. Several concurrent TTs working on one object pool are synchronized by a locking approach or by other, often tool-specific techniques (e.g., timestamps). Concurrency control information can be managed in a transient way as with conventional transactions, but must be made persistent when a TT is suspended. In contrast to DTs, TTs cooperate *implicitly* by accessing a single object pool, i.e., objects released by a certain TT are immediately available to other TTs without explicit cooperation mechanisms. Recovery can be handled by aborting TTs or rolling back partially to a savepoint. Other techniques are also conceivable, e.g., semantic undo of editor operations. If a TT performs its work in main memory buffers, a system crash can lead to a major loss of data. If it works on objects by executing ATs, the recovery mechanism for ATs guarantees that the results of successfully terminated ATs are durable.

ATs realize the *ACID*-properties. They can be used to implement TTs and certain administrative operations (e.g., starting DTs/TTs, *checkout*, *checkin*). Hereby, the conventional concurrency control and recovery concepts of a database can be exploited.

These kinds of transactions represent different levels of abstraction: DTs are realized by TTs and/or ATs, TTs are realized by ATs. Each of these levels can apply a different concurrency control or recovery algorithm. For example, DTs can be synchronized by persistent locks which survive failures or system shutdowns, TTs can use transient locks and ATs can employ an optimistic approach. However, there are dependencies between these mechanisms. A TT, e.g., can only acquire a lock if the object and the appropriate persistent lock are already available in the object pool the TT works on.

4 Configurability of the Transaction Model

4.1 Motivation

As was noted in the introduction, a transaction model for design applications has to cover a huge number of complex requirements. This especially holds in the framework area where different kinds of tools and different design methodologies are applied. No single "hard-wired" transaction model is flexible enough to meet all the requirements. Thus, in accordance with the principal idea of frameworks, we propose a transaction model that permits to individually configure a lot of different characteristics.

Example: We can observe different requirements in our *ALU* project:

- Cooperation is very intensive within a subproject, e.g., the development of the *ALU* as part of a *microprocessor*, but less intensive between subprojects. Thus, the concurrency control mechanisms applied within the subproject should be more "liberal".
- Recovery can be handled differently for interactive and batch tools. While a transaction rollback is not acceptable for interactive tools, batch tools like simulators can repeat their work automatically after a rollback.
- The consistency requirements are lower if design objects are passed within a subproject (e.g., a preliminary netlist of the *ALU* is passed to the *simulator*) than if they are released for use by other projects.

The following questions that arise in this context will be discussed in this section:

1. How can transactions be configured?
2. What are the characteristics that should be configurable?
3. How do transactions with different characteristics fit together if they are applied within *one* application or framework?

4.2 Typing of Transactions

We first discuss question 1. Transactions are configured by using a typing mechanism. This mechanism works similar to an object-oriented approach: It is possible to define **transaction types** and to specialize them by an inheritance concept. Transactions are instances of transaction types. Transactions of different transaction types can be combined in *one* transaction system in order to fulfill different requirements of applications. Similar to object-oriented class libraries, transaction type libraries can be built in advance as part of a framework and can be used and/or refined by users of the framework or by tool developers. Transaction types can be defined in a language resembling an object-oriented language or can be specified interactively. They mainly consist of methods defining the **protocols** to be used for the transaction.

Example: A transaction type defining *locking* as the concurrency control protocol can be refined into types applying *two-phase locking* and *non two-phase locking*.

4.3 Combining Transactions of Different Transaction Types

If transactions of different transaction types are used in one system (question 3) it is necessary that the protocols do not conflict. For example, if one transaction locks an object and another one accesses the object in an optimistic way an undesirable behaviour may result. Thus, an arbitrary combination of protocols is impossible.

The concept of object pools offers the possibility to define for each local workspace an individual protocol. Transactions accessing a single object pool have to use the same or at least "compatible" protocols. However, different object pools employ different protocols. Since an object pool is related to exactly one DT, typing of DTs assigns protocols to object pools.

For space reasons, we will restrict the following discussion to DTs and omit typing of TTs. Typing of ATs is not very meaningful since they are normally provided by a database system and are assumed to realize the *ACID*-properties.

As design tasks are often split into subtasks a nested DT hierarchy is built (fig. 3). The type of a certain DT defines the protocols to be used for the object pool of the DT. However, two rules have to be obeyed to make this concept work correctly.

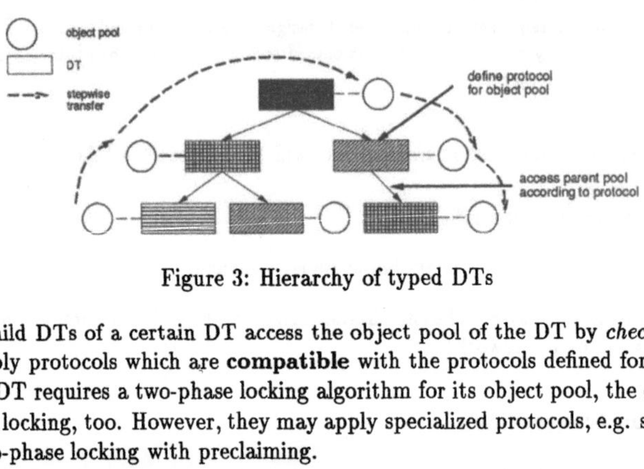

Figure 3: Hierarchy of typed DTs

First, when child DTs of a certain DT access the object pool of the DT by *checkout* or *checkin* they must apply protocols which are **compatible** with the protocols defined for that pool. For example, if a DT requires a two-phase locking algorithm for its object pool, the child DTs must use two-phase locking, too. However, they may apply specialized protocols, e.g. strict two-phase locking or two-phase locking with preclaiming.

Second, the transfer of objects between arbitrary DTs within the hierarchy has to be done **stepwise**, i.e., objects may only be transferred directly to a child DT or to the parent DT. This is important since each DT must check its protocols and may forbid certain operations. If objects were transferred directly as in the conventional nested transaction approach [Mos85] inconsistencies could result because protocols could be circumvented. The conventional approach works well only with strict two-phase locking but does not support arbitrary protocols. In particular, it does not support different protocols within one hierarchy.

Because of the stepwise transfer the subtree rooted at a certain DT builds a **sphere** with well-defined characteristics. DTs outside this sphere are not influenced by the protocols used within the sphere. This is important in order to control the information flow within the hierarchy.

We now describe in more detail some characteristics that can be configured (question 2). The list is not exhaustive, i.e., additional features can be defined. Of course, some characteristics are strongly related which means that they cannot be configured independently.

4.4 Concurrency Control Protocols

Concurrency control is important for two reasons: First, the concurrency control mechanism determines the degree of parallelism and thereby strongly influences the efficiency of the system. Second, it is a prerequisite to perform work in a cooperative way. Concurrency control requires the definition of a protocol. Locking protocols are the best-suited alternative. However, by using special lock modes we can also support a more liberal approach where objects are modified concurrently by using different copies and the changes are merged later on. To specify locking protocols two aspects must be defined:

- **Locking rules**, e.g.,
  - two-phase locking
    In this case, a DT may not acquire a new lock after it has released a lock. Two-phase locking assures the serializability property.
  - strictness
    A DT may not release any lock before it commits. This option avoids cascading rollbacks and is thus important for recovery.
  - preclaiming
    A DT must acquire all its locks at its beginning. This option avoids deadlocks.
- **Lock modes and their compatibility**
  A lock generally has two effects [Unl94]: First, the DT holding the lock is granted the right to perform certain operations (called the *internal effect*). Second, a lock specifies which operations competing DTs are allowed to perform (called the *external effect*). The distinction between these two effects enables the specification of lock modes in a very flexible way. A classical exclusive lock can be specified by an internal effect granting the right to perform all operations and an external effect granting no right at all. However, it is also possible to specify a lock where the holder may modify an object and competitors may still read it or derive new versions from it. In particular, it is possible to specify locks permitting parallel updates of different copies of a single object that are merged later on.

4.5 Cooperation Protocols

Cooperation between transactions can take place in different ways, e.g., by exchanging design objects, delegating work, sending notifications or working on common data. Cooperation must be enabled by the concurrency control protocol. Traditional protocols for *ACID*-transactions prevent cooperation, i.e., they force transactions to work in an isolated way. This is not acceptable for design environments where design is often performed in a cooperative way. The main problem with cooperation is that it allows information to flow between transactions even if it is still preliminary. Thus, it can be difficult to assure consistency and to minimize the effects of recovery actions. To deal with this problem, it should be possible to define which kinds of cooperation are permitted in a certain situation. Cooperation between DTs occurs by explicit operations based on *checkout/checkin*. We distinguish between the following cases:

- **Transferring objects**
  In this case, a DT passes a (possibly preliminary) object to another DT. The locking protocol must not be strict and – if objects are transferred back and forth between DTs – not two-phase.

- **Cooperation based on multiple copies**
  DTs check out objects into their pools and work on these copies concurrently. This is possible if appropriate lock modes are offered. Two aspects have to be configured here: First, it must be defined what happens if a DT has produced a new version of an object that is also of interest for other DTs (e.g., rereading the object, notifying the user). Second, it must be defined what happens when different updates of the same object are checked in (e.g., merging the updates).

4.6 Recovery Protocols

Recovery is the reaction of the transaction system to certain kinds of errors like system crashes, program errors or errors caused by a user. The traditional way to perform recovery, i.e. the rollback of transactions, is not flexible enough for design environments. A major problem for the recovery of DTs is cooperation: By exchanging objects, inconsistent information may be spread throughout the hierarchy which can lead to cascading recovery in case of a failure. For the configuration of recovery we apply three mechanisms:

- **Savepointing,** i.e., specifying when a savepoint is set.
- **Definition of recovery actions**
  It is possible to define how a DT reacts to failures. We distinguish between the following alternatives[2]:
  - The DT rolls back completely or – if possible – to the last savepoint (*partial rollback*).
  - The DT only rolls back the changes to the object that is affected by the failure (*selective rollback*).
  - The user or the application is notified and can perform manual or programmed recovery actions.
- **Prevention of cascading recovery**
  By restricting the flow of information in advance, it is possible to prevent cascading recovery actions. We can apply one of the following mechanisms:
  - The DT uses a strict locking protocol. In this case, no preliminary objects may be released and no other DTs may be affected by a recovery action.
  - Whenever the DT releases an object, it commits its changes to the object, i.e., it waives its right to roll back its changes later (*selective commit*).

4.7 Consistency Specifications

A main goal of transactions is to guarantee **consistency** in case of concurrency and failures. In the conventional model, consistency is defined implicitly, i.e., each transaction is assumed to be consistency-preserving. In a design environment, we should be able to define consistency in a more flexible, application-specific way and should support different degrees of consistency.

Consistency can be specified by defining a **control flow** or by defining **properties** of design objects. The first approach is used in several transaction models (e.g., [NRZ92, WR92]) and

[2] An interesting alternative not discussed here is semantic recovery by compensation [KLS90].

by design flow managers [KM92]. It fits well with our model in that control flow information is assigned to DT types. The second approach works as follows: In order to specify consistency, we use the *feature* concept [Kae91]. A feature is an arbitrary property of a design object that can have a (possibly ordered) number of values. A set of features defines the consistency state of a design object. We can use features in order to specify consistency requirements when objects are transferred. First, it is possible to specify the minimum consistency of an object that is checked out by a DT. In this way, the DT can guard itself against objects with an insufficient degree of consistency. Analogously, it is possible to specify the minimum consistency of an object that is checked in by a DT. In this way, the DT assures that is does not release objects with an insufficient degree of consistency.

Since *checkout* and *checkin* are also used for cooperation, it is possible to specify whether cooperation is permitted for objects with a certain degree of consistency.

4.8 Application Example

To illustrate our approach, we show some type definitions suitable for the DTs in our example.

Example:

- Concurrency control characteristics
  The objects in the pool of the *ALU Development* DT are versioned in a linear order. Concurrency control is done by non-two-phase locking. Valid lock modes are *share* for reading a version and *derive* for deriving a new version. In the lock compatibility matrix, *share* is compatible with *share* and *derive*, but *derive* is not compatible with itself.
- Cooperation characteristics
  The *ALU Development* DT may transfer (possibly preliminary) versions of netlists to the *ALU Simulation* DT (only for reading).
- Recovery characteristics
  The *ALU Development* DT creates an automatic savepoint for an object after a TT has modified it. The *ALU Simulation* DT rereads a netlist if it has been invalidated by a recovery action and restarts the simulation.
- Consistency characteristics
  The *ALU Development* DT may *checkin* only objects for which the simulation with certain test patterns was successful.

5 Conclusion

The transaction model sketched in this paper distinguishes itself by a high degree of configurability. This is achieved by defining transaction types using certain transaction primitives and combining them into a heterogeneous transaction system. In this way, the model is superior to other approaches and is open to model most of their features. Since the specification of transaction types is a complex task requiring knowledge in this area, a framework should provide a set of general purpose transaction types for tool developers or framework users. If necessary, a framework administrator can build new or refine existing transaction types.

An implementation of the model requires a specification mechanism for transaction types. There are two main possibilities: First, a language can be defined that should resemble an object-oriented language or be an extension of it. Second, transaction types can be defined interactively.

In our prototype based on the *JESSI Common Framework* [Ste92] we have chosen the interactive approach. However, we plan to design a suitable language because this seems more flexible.

There are some extensions that can improve our approach. First, it should be possible to use information from the data schema in order to specify protocols for certain types of objects. This leads to an integration of the language for specifying transaction types with the data definition language. Second, there should be a rule mechanism permitting the definition of automatic reactions in case of synchronization conflicts, errors or consistency violations and supporting certain forms of cooperation. We are currently investigating these topics.

References

[BS94] A. Bredenfeld and H. Streibel. Report and Demonstration of first Prototype of Complex Transaction Model. Technical Report JCF/GMD/006-01/30-Jun-94, Jessi-Common-Frame (ESPRIT Project 7364), June 1994.

[Gra81] J. Gray. The Transaction Concept: Virtues and Limitations. In *Proc. VLDB*, pages 144–154, September 1981.

[HR83] T. Härder and A. Reuter. Principles of Transaction-Oriented Database Recovery. *ACM Computing Surveys*, 15(4):287–317, December 1983.

[Ioc89] C. Iochpe. Database Recovery in the Design Environment: Requirement Analysis and Performance Evaluation. Dissertation, Universität Karlsruhe, 1989.

[Kae91] W. Kaefer. A Framework for Version-based Cooperation Control. In *Proc. 2nd Intl. Symp. on Database Systems for Advanced Applications (DASFAA)*, April 1991.

[KKB88] H.F. Korth, W. Kim, and F. Bancilhon. On Long-Duration CAD Transactions. *Information Sciences*, 46:73–107, 1988.

[KLMP84] W. Kim, R. Lorie, D. McNabb, and W. Plouffe. A Transaction Mechanism for Engineering Design Databases. In *Proc. VLDB*, pages 355–362, August 1984.

[KLS90] H.F. Korth, E. Levy, and A. Silberschatz. A Formal Approach to Recovery by Compensating Transactions. In *Proc. Conf. on Very Large Data Bases*, pages 95–106, August 1990.

[KM92] T. Kathöfer and J. Miller. The JESSI-COMMON-FRAME Project – Sub-project Development –. In *Proc. 3rd IFIP Workshop on Electronic Design Automation Frameworks*, pages 253–269, March 1992.

[KSUW85] P. Klahold, G. Schlageter, R. Unland, and W. Wilkes. A Transaction Model Supporting Complex Applications in Integrated Information Systems. In *Proc. ACM SIGMOD Conf. on Management of Data*, pages 388–401, May 1985.

[Mos85] J.E.B. Moss. *Nested Transactions - An Approach to Reliable Distributed Computing*. MIT Press, 1985.

[MUZ94] A. Meckenstock, R. Unland, and D. Zimmer. Flexible Support of Cooperative Design Environments by a Transaction Toolkit (in German). In *Proc. STAK '94*, pages 9–26, March 1994.

[NRZ92] M.H. Nodine, S. Ramaswamy, and S.B. Zdonik. A Cooperative Transaction Model for Design Databases. In A.K. Elmagarmid, editor, *Database Transaction Models for Advanced Applications*, pages 53–85. Morgan Kaufmann, 1992.

[RMH+94] N. Ritter, B. Mitschang, T. Härder, M. Gesmann, and H. Schöning. Capturing Design Dynamics – The Concord Approach. In *Proc. IEEE Data Engineering*, February 1994.

[Ste92] B. Steinmüller. The JESSI-COMMON-FRAME Project - A Project Overview. In *Proc. 3rd IFIP Workshop on Electronic Design Automation Frameworks*, pages 227–238, March 1992.

[Unl94] R. Unland. Control of Collaboration within Intelligent and Cooperative Information Systems. In *Proc. CoopIS-94*, May 1994.

[US92] R. Unland and G. Schlageter. A Transaction Manager Development Facility for Non Standard Database Systems. In A.K. Elmagarmid, editor, *Database Transaction Models for Advanced Applications*, pages 399–466. Morgan Kaufmann, 1992.

[WR92] H. Wächter and A. Reuter. The ConTract Model. In A.K. Elmagarmid, editor, *Database Transaction Models for Advanced Applications*, pages 219–263. Morgan Kaufmann, 1992.

PART TWO

Data Management and Tool Integration

3

Configuration Management in the STAR Framework *

Helena G. Ribeiro, Flávio R. Wagner, Lia G. Golendziner

Universidade Federal do Rio Grande do Sul, Instituto de Informática
Caixa Postal 15064, 91501-970 Porto Alegre RS, Brazil
e-mail {hgraz,flavio,lia}@inf.ufrgs.br

Abstract

Configuration management is an essential service to be provided by electronic design automation frameworks. Besides conventional static and dynamic configurations, normally offered by most systems, the STAR framework supports specialized facilities, such as open configurations (as in VHDL), the automatic resolution of open and dynamic configuration through expressions on object properties, and the manual resolution of these configurations by means of a graphical-interactive database browser. Unlike other systems, the STAR configuration manager is a separate framework module which the final user can directly reach through the main user interface. Furthermore, the STAR configuration management mechanisms respond to novel requirements, imposed by powerful versioning services.

1 Introduction

Configuration management is an important service that must be available in electronic design automation frameworks. Design objects to be handled in these frameworks have a complex and hierarchical structure, with objects composed of other sub-objects. The design of an object is an evolutive process, and in this context a set of versions can represent the same object in different moments of the process. When a single representation of a composite object is necessary to be submited to a design tool, a selection of versions for its sub-objects must be performed. This set of selected versions that define a given representation of the object is called an object configuration. The process of selecting versions for the sub-objects can be very long and difficult, because the sub-objects can be composed in turn of other sub-objects, and each sub-object in this hierarchy can have many versions. The user must select versions that are compatible with each other and

*This work was partially supported by CNPq and CAPES.

that best fit to given requirements. A configuration manager must provide resources to help designers establish and maintain object configurations.

Three types of configurations are defined in the literature: *static* configurations [1] associate a complete reference (object and version) to each sub-object; *dynamic* configurations [1] associate partial, incomplete references to the sub-objects; and *open* configurations [2] leave the references completely undefined. Open and dynamic configurations must be *resolved* to a complete reference before the object may be used by certain design tools.

There are many proposals of configuration management mechanisms in the literature (for instance [1, 3, 4, 5, 6]). These mechanisms support static, dynamic, and open configurations in different ways and offer different resources for the resolution of dynamic and open configurations.

This paper presents a configuration manager for the STAR framework [7], which is under development at the University of Rio Grande do Sul at Porto Alegre, Brazil. The STAR framework is based on a semantic data model that offers specialized, combined mechanisms to represent concrete design data and to manage the various representations created for each design object along the design evolution (alternatives, views and revisions). On top of the data model sits a version manager, which is responsible for revision control. The STAR configuration manager [8] supports static, dynamic, and open configurations. The hierarchical organization of versions in the STAR data model imposes some very specific requirements on the definition and resolution of configurations. Dynamic and open configurations can be resolved in a manual, automatic, or semi-automatic way, whereby versions may be selected according to desired object qualities. An interactive database browser and a special selection language are available for supporting version selection. After defined, configurations can be stored as database objects that persist in the framework and thus reused.

This paper is organized as follows. Section 2 reviews the main features of the STAR data model and of the STAR version management mechanisms. Section 3 defines configurations in the context of the STAR framework, while Section 4 presents the configuration management mechanisms. A comparison between the STAR mechanisms and those available on other systems can be found in Section 5. Section 6 concludes with final remarks.

2 The STAR framework

The STAR data model [9] provides a flexible management of the various representations created along the various dimensions of the design evolution (alternatives, views and revisions). This feature allows the system to implement, according to user- or methodology-defined criteria, conceptual schemata that are specialized for representing the design evolution of each object type.

As shown in Figure 1, each *Design* object gathers an arbitrary number of *ViewGroups* and *Views*. The ViewGroups may in turn gather, according to application-defined criteria, any number of other ViewGroups and Views, building a tree-like hierarchical *object schema*. Three types of Views are supported: *HDL Views*, for behavioral descriptions, *MHD Views* (Modular Hierarchical Description), for structural descriptions, and *Layout*

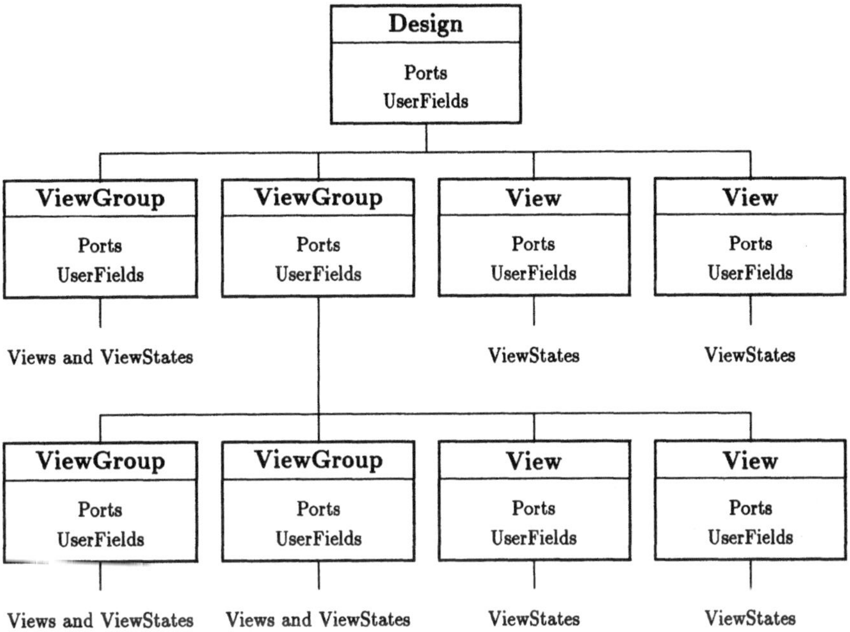

Figure 1: The STAR data model

Views, for geometric descriptions. In all View types, objects can be described as a composition of sub-objects that are instances of other objects, but only MHD Views handle the exact interconnections among the sub-objects.

ViewGroups can be used, for instance, to build a hierarchy of design decisions, where alternatives from a given design state are appended to the ViewGroup which corresponds to this state. The advantages and generality of this schema are stressed elsewhere [9, 10].

The object schema is a generalization hierarchy. Each node has properties that may be inherited by its descendant nodes. Not only the existence of an attribute is transferred to the descendant nodes, but also its value, when defined. *Ports* and *UserFields* (user-defined attributes) are among these inherited attributes. Therefore, they may be defined at any level of the object schema hierarchy.

Version management in the STAR framework [9] is supported by two different mechanisms, at two different levels. At a conceptual level, the user or the design methodology may define a particular object schema for each design object so as to organize design views and alternatives according to a given strategy. This allows the user to apply a methodology control which is highly tuned to the design of each object [7].

At a lower level, the system offers automatic mechanisms for handling the various revisions that are created for each design representation along the time axis. There are two revision mechanisms. Firstly, to each View (i.e., each leaf of the object schema) an acyclic graph of ViewStates is appended. ViewStates contain the real design data that correspond to the various design representations, such as layouts, HDL descriptions, structural decompositions, and so on. ViewStates have an associated status, representing their design stage. Possible status values are *in progress*, *stable*, and *consolidated*. Another mechanism allows the sequential versioning of the other nodes of the object schema (Design, ViewGroup, and View), according to changes made to attributes (Ports and UserFields) defined at these nodes. The system maintains the correspondence between ViewStates and versions of their ascendant nodes, thus linking each ViewState to the inherited attributes that were valid at the time of its creation.

3 Configurations in the STAR framework

In the STAR framework, a *configuration* is defined for each ViewState as a selection of a particular version of a particular design object for each sub-object within this ViewState. The three basic types of configurations already defined in the introduction (static, dynamic, and open configurations) are supported by the framework. Open and dynamic configurations must be *resolved* to a complete reference before the ViewState is used by certain design tools.

In the STAR context, the general definitions of static, dynamic, and open configurations must be refined. A static configuration selects, for each sub-object, a complete reference Design - ViewGroup - ... - ViewGroup - View - ViewState. Dynamic configurations select partial paths, containing at least a reference to a Design, but not reaching the ViewState level of the schemata of the referenced objects. The references may thus reach Design, ViewGroup, or View nodes of the schemata. In fact, in a general sense STAR configurations are always intrinsically dynamic, since the system always selects the *current* version of the Design, ViewGroup, and View nodes of the object schemata.

In order to implement open configurations, the STAR framework supports the definition of local *Components* within a given ViewState, in an approach which is similar to the VHDL language. Components are defined through their interface and parameters, and may be instantiated as sub-objects (*DesignInstances*, in the STAR terminology). Binding of Components to other design objects may be done later, when the configuration is resolved, although a reference (a complete or partial one) may be already established within the ViewState which contains the Component. Components are a generalization of DesignInstances: when there are many sub-objects of the same type, they make reference to one Component, which instantiates the common object. The same Component can be referenced by many DesignInstances.

A configuration of a design object X is presented in Figure 2. The configuration is attached to the representation defined by the path ViewGroup XVG1 - View XV1 - ViewState XVS2 in the object schema of X. XVS2 is composed of two sub-objects, the DesignInstances DI1 and DI2. DI1 references the design object MD, and the user selected the path MD - MV1 - MVS1 in the object schema of MD, while DI2 references the design object N, and the user selected the path ND - NVG1 - NV1 - NVS1 in its object schema. For each selected node (Design, ViewGroup, View) in these paths, the system automatically chooses the current revision. This is represented in the figure by hachured objects over white objects.

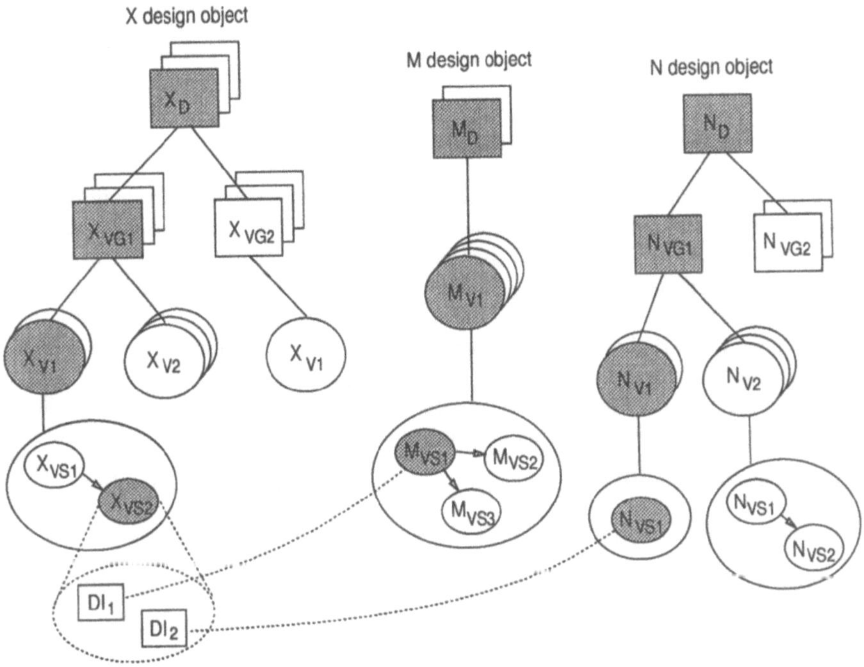

Figure 2: An example of a configuration

In VHDL, components are a mandatory support for the definition of sub-objects, even in the case of static configurations. The STAR framework offers two options: DesignInstances may instantiate either Components or directly other design objects (reaching the level of Designs, ViewGroups, Views, or ViewStates). The second alternative is more convenient for static configurations, when the object contains a small number of similar sub-objects, something that is likely to occur at more abstract design levels.

Each DesignInstance and Component has a *reference attribute* to instantiate design objects. This attribute can be defined (partial or complete references) or left undefined (no references). As in VHDL, in STAR an open or dynamic configuration may be resolved through a separate database object, called *ConfigurationBody*, which is non-versionable. ConfigurationBodies are bound to ViewStates, and one ViewState can have many ConfigurationBodies attached to it.

A ConfigurationBody completes the partial references down to ViewStates of the schemata of the objects that have been selected for the sub-objects of an object X. ConfigurationBodies may also reference already established ConfigurationBodies of these ViewStates, in a hierarchical fashion. ConfigurationBodies have a status (in progress, stable, or consolidated) which depends on the status of the referenced ViewStates.

ConfigurationBodies do not need to complete the references of all sub-objects of the object X. They may be use to specify configurations for a subset of these sub-objects, so that a complete configuration of X may result from a combination of various ConfigurationBodies.

4 Configuration manager

The *configuration manager* is a special module of the STAR framework. It is built on top of the basic data handling system and of the version manager and may be directly accessed through the framework cockpit [11]. Its goal is to support the resolution of dynamic and open configurations both through an interactive user interface and an application programming interface. The configuration manager allows the user to combine various partial ConfigurationBodies, that cover a subset of the sub-objects of the target design object, with interactive selections for the remaining sub-objects.

The configuration manager offers a repertory of operations to make possible the definition, manipulation, and resolution of configurations:

- resolve a configuration and optionally save the result in a ConfigurationBody;
- create a ConfigurationBody;
- modify a ConfigurationBody, by changing, adding, or deleting referenced objects;
- remove a ConfigurationBody, if it is not being referenced by another configuration;
- search for the existing ConfigurationsBodies for a given ViewState; and
- copy a ConfigurationBody, either to modify the copy without affecting the original ConfigurationBody or to define the same configuration for a new ViewState.

The configuration manager supports three different resolution modes: *manual*, *automatic*, and *semi-automatic.*

The manual mode is supported by an interactive database browser. The user initially selects a ViewState for which the configuration is to be resolved. For each DesignInstance of this ViewState, the user interactively selects a particular leaf of the object schema of the referenced object. This process is repetead for each ViewState selected along the object composition hierarchy. If there are ConfigurationBodies bound to a given ViewState in this hierarchy, the user can select one of them in order to avoid building a complete configuration for the whole sub-hierarchy below this ViewState.

In the automatic mode, the user chooses both the initial ViewState and one of two possible selection criteria: *current* or *most recent* ViewState. This criterion is used by the configuration manager to automatically select ViewStates for the sub-objects in the object compostion hierarchy. The current or most recent ViewState of each design object is pointed by a cursor which can be interactively moved via the browser. The automatic mode with selection of the current ViewState is the default resolution mode.

In the semi-automatic mode, the resolution is supported by a selection language that permits the expression of user-defined criteria based on desired object qualities.

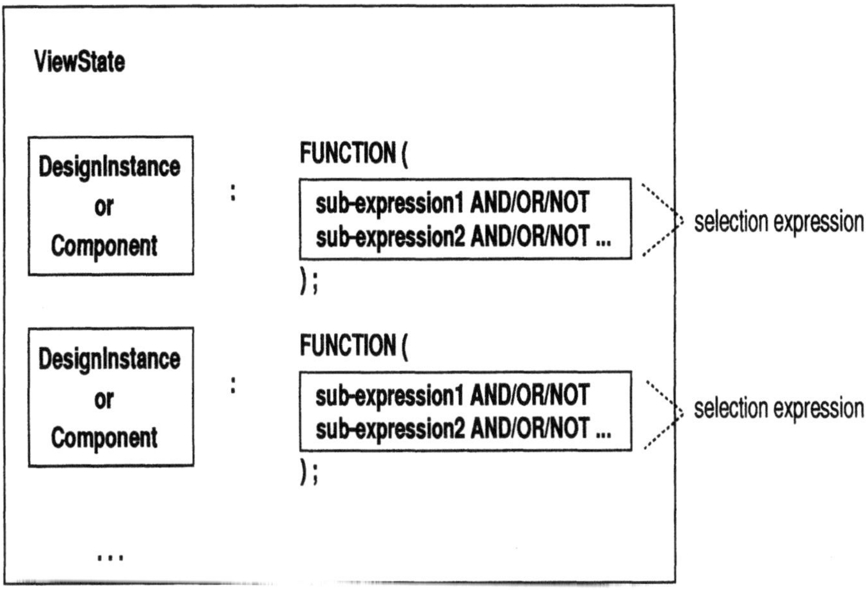

Figure 3: An example of use of the selection language

The selection language (see example in Figure 3) allows the user to specify *configuration expressions* for the Components (or DesignInstances) of the ViewState for which a dynamic configuration is to be resolved. The expressions may be attached to each Component (or DesignInstance), to groups of Components (or DesignInstances), or to all Components (or DesignInstances) of the ViewState.

An expression is composed of a set of sub-expressions. Each sub-expression specifies conditions on properties the ViewStates must show in order to be selected. The evaluation of each sub-expression restricts the set of candidate ViewStates. The simultaneous evaluation of all sub-expressions results in a final set of candidate ViewStates that show all the desired properties. In order to restrict this final set to a single ViewState, the user must either define a conflict resolution function (which specifies the selection of either the current or the most recent ViewState), or switch to the manual selection mode, or redefine the configuration expression. If the final set is empty, the resolution fails.

The sub-expressions are built as logical connections (through AND, OR, and NOT operators) of *logical factors*. Factors are expressed by the following special operators:

- *relational operators*, which compare object attributes to constant values;
- MAX (or MIN), which selects the version with the maximum (or minimum) value of an attribute;
- *existencial operators*, which ask for the existence of attributes (possibly inherited ones);

- LAST (or FIRST), which selects the last (or first) version of a Design, ViewGroup, or View; and
- CURRENT, which selects all current ViewStates associated with a Design, ViewGroup, or View.

5 Related work and comparison

As in the SDE environment [5, 12], the STAR configuration manager offers configuration expressions as a means of specifying user-defined constraints to solve dynamic and open configurations. The SDE environment supports four types of constraints:

- *performance* constraints specify restrictions on attributes like size, delay, power consumption, and latency;
- *selection* constraints are restrictions on the selection of cells as components;
- *environment* constraints specify the environment where a cell can be used as a component of a larger design (e.g related to a cell fanout); and
- *relativity* constraints are dependencies of cells on other cells;

Other systems, such as the OCT manager [3] and the Version Server [1] use *workspaces* (called *layers* in the Version Server) to organize versions that may be used together in a meaningful configuration. In these systems, the user is directly responsible for assigning versions to workspaces (or layers). These mechanisms are directly comparable to the *selection* and *relativity* constraints of the SDE environment. They do not allow the expression of complex queries to solve configurations.

The *selection language* of the STAR framework, in turn, allows for the specification of expressions that are directly comparable to the *performance* and *environment* constraints of the SDE environment. The assignment of adequate attributes to the various nodes of the STAR object schemata allows for an "emulation" of the SDE *selection* and *relativity* constraint types.

In the SDE environment, constraints are embedded into a VHDL extension and are thus expressed within the design object descriptions. In the STAR framework, the configuration manager is a separate module offering a general-purpose service for resolving dynamic configurations. Design tools and descriptions are not affected by this service. If desired, design tools may access it through the API of the configuration manager.

Configuration management in the STAR framework has very particular requirements. The hierarchical object schemata and the possibility of assigning Ports and other general-purpose attributes to any of their nodes make dynamic configurations and their resolution more complex, but give an extra modeling flexibility.

6 Final remarks

The configuration manager is a separate module of the STAR framework, which offers integrated, extensive, and flexible mechanisms to define, solve, and handle design object configurations. These mechanisms are implemented through both an interactive user interface and an application programming interface. The main features of the configuration manager are:

- support to static, dynamic and open configurations, so as to define configurations according to the current development of the design object or to the abstraction level which is desired for a given object representation;
- the definition and storage of ConfigurationBodies and its reutilization, also in a hierarchical way, within other configurations;
- the resolution of dynamic and open configurations in one of three different modes (manual, automatic, and semi-automatic);
- a selection language that allows the definition of configuration expressions to select objects according to object features, so as to establish different semantic configuration constraints; and
- an interactive selection of object representations, by means of a navigation in the object schemata via an interactive database browser.

References

[1] R.H.Katz et al. Design version management. *IEEE Design & Test of Computers*, February 1987.

[2] IEEE. *IEEE Standard VHDL Language Reference Manual*, 1988.

[3] M.Silva et al. Protection and versioning for OCT. In *26th Design Automation Conference*. ACM/IEEE, 1989.

[4] E.Siepmann. A data management interface as part of the framework of an integrated VLSI design system. In *Intern.l Conference on Computer Design*. IEEE, 1989.

[5] M.J.Chung and S.Kim. The configuration management for version control in an object-oriented VHDL design environment. In *Intern. Conference on Computer Aided Design*. IEEE, 1991.

[6] S.Banks et al. A configuration management system in a data management framework. In *28th Design Automation Conference*. ACM/IEEE, 1991.

[7] F.R.Wagner, L.G.Golendziner, and M.R.Fornari. A tightly coupled approach to design and data management. In *EURO-DAC*. 1994.

[8] H.G.Ribeiro. A Configuration Manager for the STAR Framework. Master thesis, UFRGS, Porto Alegre, 1993.

[9] F.R.Wagner et al. Design version management in the STAR framework. In *3rd IFIP Intern. Workshop on EDA Frameworks*. North-Holland, 1992.

[10] F.R.Wagner and A.H.Viegas de Lima. Design version management in the GARDEN framework. In *28th Design Automation Conference.* ACM/IEEE, 1991.

[11] R.Mello, L.G.Golendziner, and F.R.Wagner. The visual interface of the STAR framework. Also presented at the *4th IFIP Intern. Working Conference on EDA Frameworks*, 1994.

[12] S.Kim and M.J.Chung. A constraint-driven approach to configuration binding in an object-oriented VHDL-CAD system. In *10th IFIP International Symposium on Computer Hardware Description Languages and Their Applications.* North-Holland, 1991.

4

AutoCap: An Automatic Tool Encapsulator[1]

João Câmara *, Helena Sarmento **

R. Alves Redol, 9, 1000 LISBOA, Portugal

Phone +351 1 3100353 Fax: + 351 1525843 email jac@olivia.inesc.pt

* INESC ** IST / INESC

1. Abstract

CAD frameworks often have to be updated by incorporating new CAD tools. Encapsulation is the tools' incorporating technicality most often used. Once tools have been incorporated into a framework, the designer access to them may be unified and simplified.

The present work describes work recently developed, within the PACE framework, to address tool encapsulation and tool invocation.

2. Introduction

The dynamic nature of CAD systems implies the ability to easily modify them. These modifications often include adding new tools to the framework (or removing outdated ones).

Tools can be incorporated into a CAD system in two ways: integration and encapsulation. Integrating a tool means tightly connecting it to a CAD system by writing or modifying the tool to use framework services directly to get data and other information it needs to be executed. Encapsulation means loosely interfacing a tool to a CAD system trough a layer separated from the tool. Integrating a tool requires the tool's source code to be available and usually represents a substantial investment of time and expertise. On the other hand, encapsulating a tool is usually less time consuming and no modification of the tool's source code is required. This is

[1] Part of this work was done under ESPRIT project SPRITE.

why, when incorporating tools into a framework, encapsulation is more often used.

Nowadays CAD systems need to incorporate a large number of different tools, each one with its own particular, often unique, requirements for proper execution. The need to remember the correct tool location, command syntax and arguments necessary to invoke each tool often leads to errors and decreased productivity. The development of CAD frameworks, which are software infrastructures that provide a common operating environment for CAD tools, is very important since they offer a number of services that ease the task of managing and updating CAD systems. Due to the dynamic nature of CAD systems, the easiness to incorporate a tool into a framework is an important factor to be considered when evaluating a framework.

Once a tool has been incorporated into a framework, tool management, one of the services offered by CAD frameworks, frees the designer from the need to remember the correct location, command syntax and arguments necessary to invoke each tool.

This paper describes recent work developed within the PACE framework [Sarmento 90] to address two related problems: easiness of incorporating a tool into a CAD framework and easiness of invoking tools from a CAD framework. Section 3 refers related work developed within several CAD frameworks. Some aspects of tool encapsulation are mentioned in section 4. Section 5 briefly describes TES, the CFI standard for tool encapsulation information. Sections 6 and 7 describe AutoCap. Finally, some conclusions are presented in section 8.

3. Related work

The problem of incorporating tools into a framework has been addressed by different frameworks in different manners.

Both ULYSSES [Bushnell 85] and CADWELD [Daniell 89] frameworks are based on a *blackboard architecture*. The *blackboard* is global database containing references to all the files related with the tools incorporated in the system. In the ULYSSES framework the characterisation of a tool is part of the blackboard. In CADWELD, tool characterisation is obtained by building a software layer that allows the tool to interact with the blackboard. Both CADWELD and ULYSSES implement a name server as part of the blackboard, thus allowing an easier invocation of the tools.

Within FALCON [Mentor 90] framework tool characterisation is done by the association of a *qualification script* to each tool. This framework includes a *Design Manager* that permits the designer to easily invoke a tool with a graphical interface quite similar to the one used in Apple Macintosh [Apple 91] computers.

CADLAB [TIDL 90] framework offers an object oriented language, TIDL[2], to describe the design environment. The tool integrator uses this language to define the objects needed to encapsulate the tool. TIDL automatically generates de dialogue boxes that allow the designer to enter the tool's parameter list.

Within DESIGN FRAMEWORK II [DFII], the tool integrator writes a program, using the SKILL language, to wrap the tool thus permitting the management of the tool within the framework. Once encapsulated, tools are invoked by invoking the encapsulation function, which can be associated to the call back function of a menu item.

---

[2] Tool Integration and Description Language

4. Tool Encapsulation

As referred, the evolution of a CAD system often requires adding tools to the framework. The existence of tool encapsulation mechanisms is therefore of great importance.

Tool encapsulating requires both a tool description language to characterise the tool and a mechanism capable of interpreting a tool description. Although, theoretically, any description language may be used, the language should obviously be concise and have an adequate expressive power. It is also advantageous that the language has a human and computer readable form, since this allows the information contained in a tool description to be used in either a manual or automatic creation of a tool encapsulation.

Aiming at the implementation of a standard for a format for tool encapsulation information, CFI has developed TES (Tool Encapsulation Specification) [CFI 92], which we briefly describe in the next section.

5. TES

The current version of TES includes general tool information, description of all the tool arguments, description of all data associated with the tool function, description of the syntax for a valid command line to initiate the tool function and description of possible result codes. TES also includes a mechanism that permits to add arbitrary name/value pairs to a description construct, thus providing with an extension capability the current version of TES.

We have adopted TES as our description language and developed AutoCap (Automatic Encapsulator), a tool that interprets the information contained in a TES description and uses it to automate the encapsulation of the tool described.

6. AutoCap

AutoCap reads a TES file and maps it to an internal representation structure. This structure is complex enough to deal with all types of information needed and general enough to be usable in the construction of any type of user interface (graphical, textual or other). The internal representation structure has to be of some complexity since AutoCap needs to represent all the information contained in the TES file internally. Alternatively AutoCap could deal separately with each information construct contained in the TES file, thus allowing the use of a simpler internal structure; however this approach would not allow the global treatment of the information, necessary in the construction of a graphical interface.

The information contained in a TES file may be used in two ways:

- read the TES file once, at the time the tool is registered with the framework, extracting its information and storing it in persistent objects which are then activated by the framework when the designer invokes the tool;
- read the TES file each time the tool in invoked using an interpreter-like program to extract the information and use it to invoke the tool directly.

AutoCap uses the second approach since it is simpler and since we have verified that the TES file interpretation may be done quite quickly. This approach also permits, when the framework invokes AutoCap, to pass to AutoCap parameters that are only defined at run time.

AutoCap automatically builds a user interface which is a graphical representation of the tool arguments described in the TES file. The designer is presented with a user friendly interface, where the values required are labelled and may be of different types: integer, real, string, multiple choice[3]. The required values may be entered either by pushing buttons or by typing then onto entry fields. Figure 1 shows an example of an AutoCap interface window.

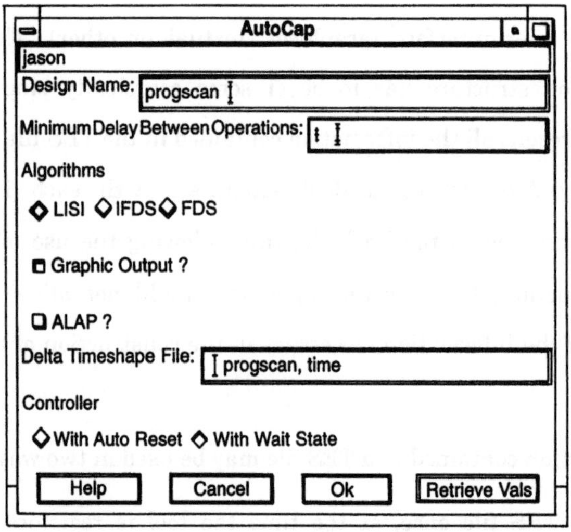

Figure 1 - AutoCap interface window

The arguments the designer is required to enter do not necessarily correspond to the arguments used in the command line to initiate the tool function: some of them may be used to somehow evaluate other values that, in turn, are used in the command line. Similarly the values inputted by the user for some arguments may correspond to different values in the command line: for instance the value *true* of a boolean argument may correspond to the value *-g* in the command line.

Ranges of acceptable values for an argument may be specified in a TES file. AutoCap checks if entered values comply the restrictions specified in the TES file, if any. In case of an error, a *popup window* is used to notify the designer.

[3]Multiple choice arguments are represented either by radio buttons (exclusive multiple choice) or by toggle buttons (inclusive multiple choice).

A TES file may contain help lines. These lines are used to supply additional information to the user concerning the enclosing description constructs. On demand, AutoCap opens a window (see figure 2) presenting this information to the user.

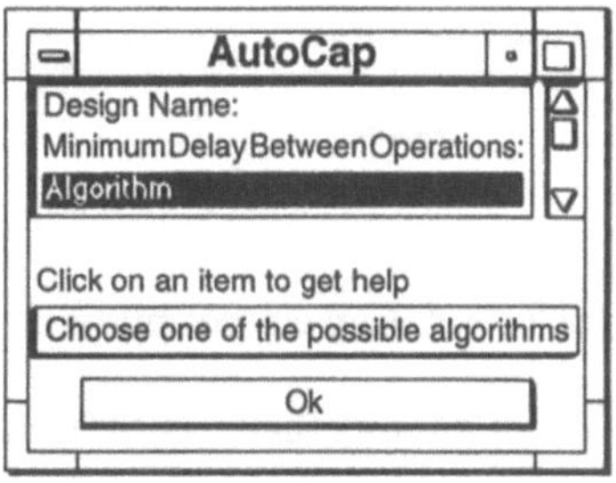

Figure 2 - Help window

When invoking a tool the designer often enters for each argument the same value he did last time the tool was invoked. This led to the existence of the *Retrieve Values* button (see figure 1) in the AutoCap interface window. By pressing this button the designer is able to retrieve the argument values he entered last time the tool was invoked.

7. Implementation

AutoCap was developed using C++[4], and it implements a set of classes that allow the storage of the tool encapsulation information. These classes are associated with the different entities the information refers to, such as tool, argument, argument list, value, concatenated value and others. Figure 3 depicts the class hierarchy, using the notation proposed in [Rumbaugh 91].

AutoCap graphical interface was built using the services of GHOST [Santos 90], a graphical server that greatly simplifies the programming effort necessary to implement the interface.

[4] Approximately 6000 lines of code were written.

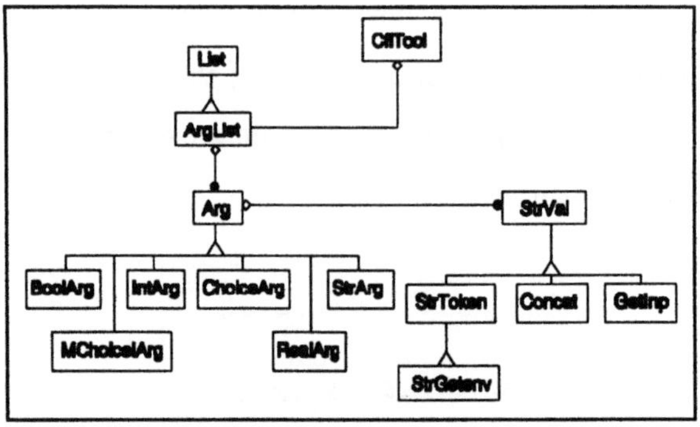

Figure 3 - AutoCap classes hierarchy

To deal with a TES file AutoCap uses the tools LEX and YACC. The use of these tools showed to be extremely useful since TES allows great flexibility in the order and number of elements that constitute each of its description constructs.

8. Conclusions

AutoCap is presently being used by DESKTOP [Martins 93], a part of the PACE framework, to invoke the encapsulated tools.

AutoCap was used to encapsulate several tools, such as ARCHITECT (a schematic diagrams visualisation tool developed at INESC) and JASON (a scheduler tool of the PHIDEO compiler developed at PHILIPS Natlab). Although there were no TES files available for these tools, these files where very easily written by the tool integrator, once he was told about the tool arguments.

AutoCap greatly simplifies the work of a tool integrator, since it only requires the existence of a TES file describing the tool. This file is supposed to be supplied by the tool vendor, so all the information the tool integrator has to add is the tool location. So, unlike the methods used by other frameworks referred in section 3, the tool integrator's work is reduced to an absolute minimum. By using a common user friendly interface, AutoCap also simplifies the designers work as far as tool invocation is concerned.

The development of AutoCap was greatly simplified by the use of GHOST, another service provided by the PACE framework. On the other hand, the existence of AutoCap was very useful in the development of BALANCE, an automatic dynamic load balancer recently offered by the PACE framework. By capturing information about all data associated with the tool function and possible result codes, AutoCap may also be very useful for flow management.

References

[Bushnell 85] M. L. Bushnell, S. W. Director, *ULYSSES - An Expert-System based VLSI Design Environment*, International Symposium on Circuits and Systems, June 1985.

[CFI 92 - a] TCC Approved Draft Proposal - CFI, *Toll Encapsulation Specification*, CAD Framework Initiative pilot release document CFI-92-P-10

[CFI 92 - b] CAD Framework Initiative - CFI, *Standards Release*, Notifications of CFI 2.0 Pilot Program, 1992.

[Daniell 89] J. Daniell and S. Director, *An Object oriented approach to CAD Tool Control Within a Design Framework*, 26th ACM/IEEE Design Automation Conference, 1989

[Martins 93] J. F. Martins, Final Prototype of a Desktop, technical report, INESC, SPRITE Esprit project, November 1993.

[Mentor 90] Mentor Graphics, *The Falcon Framework Technical Papers*, June 1990

[Rumbaugh 91] J. Rumbaugh, M. Blaha, W. Premerlani, F. Eddy and W. Lorensen, *Object-oriented Modelling and Design*,. Prentice-Hall International Editions, 1991.

[Santos 90] P. Santos, H. Sarmento and L. Vidigal, *Ghost/Spook User Interface and Process Management in the PACE Framework*, European Design Automation Conference, March 1990.

[Sarmento 90] H. Sarmento and P. Santos, *A Framework for Electronic Design Automation*, IFIP Workshop on Electronic Design Automation Frameworks, North-Holland, November 1990.

[TIDL 90] Cadlab, CADLAB Tool Integration Description Language Release 3.0 TIDL - Overview, 1990

5

Schema Evolution in the STAR Framework[1]

Miguel R. Fornari[2], Lia G. Golendziner, Flávio R. Wagner

Universidade Federal do Rio Grande do Sul, Instituto de Informática
Caixa Postal 15064, 91510-970 Porto Alegre RS, Brazil

E-mail: {miguel, lia, flavio}@inf.ufrgs.br

Abstract

The STAR data model supports the definition of object schemata, according either to some design methodology or to the designer's decision. Object schemata allow a flexible management of the various representations that are created during the design of a particular object. Object schemata can evolve or even be dynamically defined, departing from an existing object schema and making changes to it. Schema evolution facilities are a valuable support for both the definition of new design objects and design methodology management. Schema evolution is maintained through versions, so that version management is applied not only to design objects themselves, but also to object schemata. Consistency is guaranteed for schema operations, based on a set of invariant rules.

1 Introduction

Typical EDA frameworks are built upon a database management system that offers data representation facilities and basic versioning mechanisms. On top of this layer, various servers, eventually implemented as domain-neutral tools, are available. Typical servers support the management of versions, configurations (both aspects of data management), and design methodologies.

In a complex design environment, where prototyping is a common way of developing systems, there is a great need to change the schema definition [Bert91]. This process requires a special tool, incorporated to the database, to allow operations over the schema [Atki89]. Schema updates should not represent an extreme overhead and, at the end of the changes, schema and instances must be correct and consistent. There is also a need for flexibility, which tends to make this kind of operation more expensive to the system.

This work describes an innovative data definition layer that allows the creation and evolution of object schemata in the STAR framework [WGF94]. The mechanism uses object

[1]This work was partially supported by CNPq and CAPES.

[2]Currently at Universidade Luterana do Brasil (ULBRA). Caixa Postal 124, CEP 92420-280, Canoas-RS-Brazil.

versions to represent several states of the same object and to permit the return to previous states of one object in a natural way. Changes to an object schema are not done "in-place", but new versions are generated, so that a schema evolution history is kept. The main objective of this mechanism is to allow the construction of a new object schema based on an existing one. Tested object schemata can then be reused, improving design team productivity and increasing reliability of resulting systems.

The remaining of the paper is organized as follows. Section 2 introduces the STAR data model, the design methodology layer and the version model. Section 3 presents the schema definition and evolution layer. Section 4 presents a comparison of this mechanism with those existing in other frameworks and object-oriented database systems and Section 5 concludes with final remarks.

2 The STAR framework

2.1 The STAR data model

In the STAR data model, shown in Figure 1, each *Design* object gathers an arbitrary number of *ViewGroups* and *Views*. The *ViewGroups* may in turn gather, according to user- or methodology-defined criteria, any number of other *ViewGroups* and *Views*, building a tree-like hierarchical object schema.

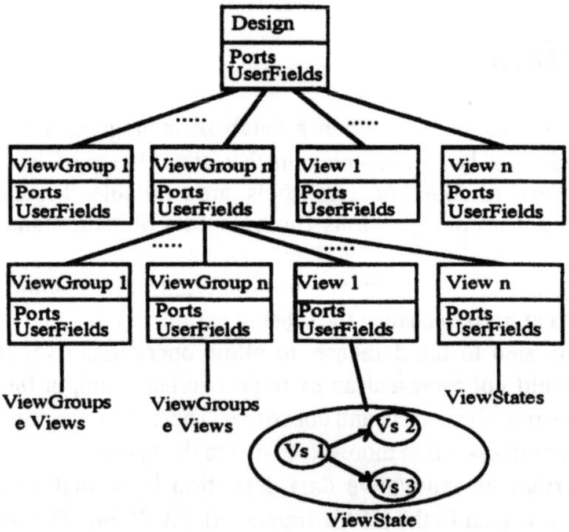

Figure 1: The STAR data model

The fact that *Views* may be defined at any level of the object schema offers an unlimited number of ways for organizing the different representations of the *Design*. Since the system does not enforce any grouping criterion, it is left to the user or to the design methodology to decide how *Views* will be organized. *Views* are of one of the types: HDL, for behavioral descriptions, MHD, for structural descriptions and *Layout*, for geometric descriptions.

The object schema is a generalization hierarchy, in that each node is an abstraction of the subtree below it. Properties defined at each node may be inherited by its descendant nodes (inheritance is optional). Inheritance occurs among instances: not only the existence of an attribute is inherited by the descendant nodes, but also its value, when defined. Inheritance may be by default, when descendant nodes may redefine attributes to more specialized domains and modify attribute values, or strict, when redefinition at descendant nodes is not possible.

The purpose of *Design*, *ViewGroup* and *View* nodes of the object schema is to organize the various representations of a *Design* object and to guarantee the consistency of the common attributes through the inheritance mechanism. Therefore, these nodes contain only the attributes to be shared by the representations they gather.

Real design data, such as structural decomposition, HDL descriptions, and layout masks, are contained in the *ViewStates*, that are revisions created for each of the *Views*.

There are three types of attributes for each node of the schema: *UserFields*, *Ports* and *Parameters*.

UserFields are user-defined object attributes, which have a name and a domain specified. The domain can be simple (character, string, integer, real, boolean) or composed (record, array, set, subset and enumeration). Inheritance is optional for *UserFields*, that is, attributes can be defined as "local" to one node and not passed down to the inheritance hierarchy. If the inheritance type is defined as default for a *UserField*, its domain can be redefined, but only making restrictions to it. The new domain must be a subset of the original one.

Ports are interface signals and may contain in turn their own user-defined attributes. *Port*'s definition specifies: a domain, that indicates the type of information contained; a direction (in, out or inout); and the number of wires. There are two types of *Ports*: *PortWires* with just one wire and *PortBundles* composed by a set of *PortWires*. *Ports* are interconnected by *Nets*.

Parameters allow the user to build generic, parameterized objects. *Parameters* have only a name and a domain. Inheritance is mandatory regarding the existence of *Ports* and *Parameters*.

The STAR data model allows the specification of relationships between objects, which is done through an object of type *Correlation*. A *Correlation* has a direction and a mode. The direction (*bi-directional, directed* or *non-directed*) indicates an existence dependency. For example, a correlation defined as A → B (A directed to B) indicates that B can only exist while A exists. The mode (*protect* or *delete*) indicates the action to be done when the removal of an object is required. If the mode is *delete,* a removal of a node causes the removal of the nodes that depend on it through *Correlations*. Otherwise (protect mode), a removal cannot be executed if there are dependent nodes. In a non-directed *Correlation*, the mode is irrelevant. *Correlations* can also have *UserFields* and a relationship criterion, for documentation purposes.

It is important to notice that an object schema in STAR is a hierarchy that presents inheritance not only for attribute definition (as occur for most of the OODBMS) but also for attribute values, which can be defined at any level of the hierarchy. Value inheritance brings up two problems into consideration: modifications in object descriptions and modifications in attribute values.

2.2 Design methodology management

A design methodology is a set of design rules that either enforce or guide the activities performed by the user, so as to obtain objects with desired properties. The definition of a design methodology in the STAR framework is based on three main principles [WGF94]: the definition of the object schema for the design objects, the specification of the task flow and the hierarchization of design strategies. The object schema has been already discussed above.

Task flow is expressed through a condition-driven model. A task is described with a 4-tuple (*name, pre-conditions, tool, post-conditions*). A task is eligible for execution when its pre-conditions hold. These conditions can express the existence of objects or properties of them, explicitly modeled as attributes in the object schema. To execute the task, the specified tool is used. Post-conditions describe the properties expected from the objects after a task is executed. Again, a set of new objects, generated by the tool can be expected as the result of the execution. If the post-conditions are not achieved, the task fails, though new object representations might have been created. A task execution is considered a long database transaction, whose effects can be undone if the user asks for. It is left to the user to select among many enabled tasks. A methodology succeeds when all its tasks have succeeded. Tasks may be executed stand-alone or within a design strategy.

Design methodologies can be organized in a hierarchical way. A new design methodology can be derived from a previous one by extending the object schema (using the schema evolution mechanism) or defining new tasks.

Task definition must be consistent with the object schemata that are known to this methodology, that is, all referenced objects and attributes must exist and attribute comparisons must be done with correct domain values.

2.3 Version management

The STAR framework provides a two-level versioning support [Wagn92]. At a conceptual level, the object schema defines *ViewGroups* and *Views* that represent different design views and alternatives, according to user or methodology control. At a lower level, revisions are automatically generated by the system, when updates are done to specific representations of the design object.

There are two revision mechanisms. First, to each View an acyclic graph of ViewStates is appended. They contain the real design data that corresponds to the various design representations (layouts, HDL, descriptions, and so on). Another mechanism allows the sequential versioning of the other nodes of the object schema (*Design*, *ViewGroup*, and *View*), due to changes made to attributes that were defined as versionable. The system maintains the correspondence between *ViewStates* and versions of ascendant nodes, thus linking each *ViewState* to the inherited attributes that were valid at the time of its creation.

Versions have an associated status, representing their design stage, which can be *in progress, stable* or *consolidated*. In progress versions can be changed or deleted. Stable versions can be deleted, but not changed. To maintain the historical sequence of versions, they are only logically deleted. Consolidated versions can not be changed nor deleted. They can only be selected and read. When a version is promoted (to stabilized or consolidated), its predecessors are also promoted to the same status.

Attributes (*UserFields, Ports* or *Parameters*) can be defined as versionable or non-versionable. A modification on a versionable attribute implies a creation of a new version, exception made for versionable attributes of in progress versions, which can be modified. The automatic revision control guarantees that when a stabilized object is modified, a new version is created as a copy of it, but with the modified values (if the modified attribute was defined as versionable). A non-versionable attribute can not be modified in any way. The new version is created with in progress status.

For all objects having versions in the object schema, there is the notion of *current version* (by default, the most recent one). The user can query old versions without changing the current one. Changing current versions is made through a selection operation, in one of the two following ways:

Partial: A version is selected from one node of the object schema and only the current version of this node is changed. Current versions of all other nodes in the object schema are not changed. It is user's responsibility to verify consistency among the version selected for this node and the others in the object schema.

Total: The user chooses either a specific version in one of the nodes or a *ViewState*. The system then changes the current version of all the ascendant nodes to the version that were valid at the creation time of the chosen version (or *ViewState*).

3 Schema Evolution

Due to the nature of the design process, an object schema may need to be dynamically modified in several ways, reflecting new specifications and user requirements, inclusion of a new tool to the environment and correction of modeling errors. In particular, schema evolution is an essential feature for supporting design methodology management in an evolving environment, where the inclusion of new tools and strategies, during the design process, may impose the incorporation of new types of object representations and new attributes to the already existing object schemata.

A mechanism for the definition and evolution of object schemata has been developed for the STAR framework. As in object-oriented databases [PeSt87, KiCh88, Deux91], this mechanism is based on *schema invariants*, which are basic conditions that must always hold to insure that the object schema is in a consistent state.

Object schemata can be created from scratch, a simple situation for the schema evolution manager, but it is highly desirable to develop an object schema from an existing one, well tested and approved. To achieve this, the first operation must be the copy of one existing schema to a new area. The inclusion of new *ViewGroups* and *Views* does not affect *ViewStates* that already exist. Modifications in nodes having associated ViewStates can also be done. In this case, new *ViewStates* have to be generated, reflecting the modification in ascendant nodes. This method of object schema definition results in an important decrease of design development time.

3.1 Schema Invariants

Invariants for the schema evolution mechanism were defined to assure database integrity and object schema correctness. The invariants for the STAR schema evolution are[3]:

[3]For simplicity, object means Design, ViewGroup or View. Instance means ViewState and attribute means UserField, Port or Parameter.

- Each object inherits properties from only one other object (single inheritance). This restriction comes from the definition of the object schema as a tree-like generalization hierarchy.
- All descendant nodes from an object have a unique name. The complete name of an object is composed by the Design name and all the descendant object names that are in the path to the mentioned object in the object schema.
- All attributes of an object have unique names. This restriction guarantees attribute identification.
- All inheritable attributes of an ascendant object are inherited. Since the first invariant assures that multiple inheritance does not exist, name conflicts do not occur.
- *UserFields* inherited by default can be redefined in descendant nodes: the domain can be redefined to more specialized domains or values can be modified. Strict inherited attributes (some *UserFields*, all *Parameters* and *Ports*) can not be redefined.
- All referenced objects in the schema are present in the database. This is the referential integrity of relational databases adapted to the STAR framework.

This invariant set assures the integrity of the database objects. However, sometimes a sequence of operations is needed to go from a consistent database state to another one. Then, a modeling transaction can be started, disabling the invariant checking until the transaction is committed. One exception is the stabilization of a version. When this operation is required, the hierarchy where the object version is included must be verified to assure that only correct versions are stabilized. This modeling transaction is typically a long transaction and must be incorporated in the mechanism of long transactions provided in the framework.

3.2 Operations

A complete set of operations for schema modification is defined to allow an easy data modeling and evolution in the STAR framework. These operations are the basis for a higher level object schema definition language and are listed below. The name of the object is in bold, its properties are in italic and the possible operations on an object are cited. Brackets indicate an option to be taken. Square brackets indicate an optional value. If not specified, a NULL value is assumed.

- **Library** *(Name)*
  Operations: Create; Delete.
- **Design** *(Name, Library)*
  Operations: Create; Delete.
- **ViewGroup** *(Name, {Design, ViewGroup}, [Criterion])*
  Operations: Create; Delete; Modify ascendant object or criterion.
- **View** *(Name, {Design, ViewGroup}, Type)*
  Operations: Create; Delete; Modify ascendant object.

- **Parameter** *(Name, Domain, Inheritable, Versionable, Object)*
  Operations: Create; Delete; Modify domain, versionable characteristic and/or inheritable characteristic.
- **UserField** *(Name, Domain, Inheritable, Versionable, Inheritance Type, Object, [Value])*
  Operations: Create; Delete; Modify domain, inheritable characteristic, versionable characteristic and/or inheritance type; Modify value; Move the *UserField* to another object.
- **Port** (*Name, Type, Object, Versionable, Direction, [Number of wires], [Domain]*)
  Operations: Create; Delete; Modify versionable characteristic, direction, number of wires and/or domain; Move the *Port* to another object.
- **Correlation** *(LeftObject, RightObject, Direction, [Mode], [Criterion])*
  Operations: Create; Delete; Modify objects, direction, mode and/or criterion.

When a new node is created, its name and its immediate ascendant must be informed. The name of the node must be unique, according to the schema invariants. When a node is removed, all its descendant nodes are removed too. For in progress versions, the design data are really removed. For stable versions, the data are maintained in the database, but just historical queries can be done on them. Consolidated versions cannot be removed. A node can change its place in the object hierarchy, moving to another ascendant object. This operation is semantically equivalent to a combination of a removal from the original node and an insertion in the new one. All descendant nodes, if they exist, are moved together.

Attributes can be inserted, removed and copied at any time. If the current version status is not in progress, a new version is derived from it, and the attribute modification is effective in this new version.

If an inheritable attribute is redefined, the inheritance mode must be verified. If this attribute redefines another inherited attribute, then the inheritance mode should have been defined as *by default*, and the redefined domain must be a subset of or equal to the inherited domain.

When the domain of a *UserField* is modified, its value should be changed to keep consistency. The user can define a special function that automatically maps the old values to values in the new domain.

The modification of an attribute from versionable to non-versionable can be done at any moment. This modification alters only the semantics of value modification of an object and does not change the object schema.

Modifications in the number of wires in a *Port* are possible just for *PortBundles*. In case of reduction of *PortWires* it is necessary to indicate *PortWires* to be removed. If the number of wires is increased, a list of new *PortWires* has to be indicated. When the direction of a *PortBundle* is modified, the direction of all *PortWires* that composed it must be modified too. Designers receive a list of *Nets* that are affected by the modification, to allow a manual correction.

ViewStates are not considered in the schema evolution manager because they are instance objects. Creation, removal and other operations on *ViewStates* are directly controlled by the data manipulation language (DML). All modifications in the object schema are reflected in the instances, i.e., the *ViewStates*. The automatic revision control generates a new version when some characteristics are modified, in a consistent way.

Correlations can be freely modified because changes on them do not violate any invariant, just modify the semantics of the delete operation.

Copy of a node is an essential operation to allow the designer to reuse a well tested and approved object schema in the development of a new design. The designer can copy a well-established design to a new area and make some modifications to obtain the necessary conceptual schema for the new design. The designer can copy just one node or the node and all its descendants.

A great number of schema evolution operations can affect the correction of already defined tasks. For example, an attribute used to express a pre-condition of a task, if removed, turns the tasks' definition incorrect. When such situation occurs, a list of incorrectly defined tasks is returned to the designer, who is responsible for making the necessary modifications for correcting the affected tasks.

4 Comparison

In the STAR framework, both the final user and the application programmer have full access to the schema evolution facilities, including removal and redefinition of nodes and attributes.

In the CADLAB framework [Gott88], for instance, the final user may only extend existing schemata, by using the TIDL language [Groe91] and recompiling the schema definition. The application programmer, in turn, may also delete attributes and object types for which no instances exist.

In the NELSIS framework [Wolf88], the database system offers a semantic data model (called OTO-D), versions, a graphical query interface, tool activation, support for design transaction and physical distribution, but there is no schema evolution facility.

The power of the schema evolution mechanism of the STAR framework can be compared to those present in object-oriented databases, considering the large number of available operations.

Invariants are used by ORION [KiCh88], O_2 [Deux91] and GemStone [PeSt87]. These invariants guarantee the structural correctness of schemata, but do not include behavioral aspects. In STAR, the mechanism that returns to the designer a list of affected tasks is similar to the mechanism that controls method modifications in O_2 and is called behavioral consistency [Zica91].

The possibility of combining versions of nodes and schema evolution presents similarities with the proposals by Kim & Chou [KiCh88] and Skarra & Zdonik [SkZd86]. However, in [KiCh88], versions of the whole schema are generated after a schema modification. In [SkZd86], versioning is done for a single class. Modifications that impact stored objects (for example, changing a domain of an attribute) must be managed by *handlers* provided by the user. In the STAR mechanism, versions are created for any node in the object schema. Version nodes are connected so that it is possible to return to previous versions of any node, keeping the correspondence among all versions of the schema and the design data (represented by the *ViewStates*).

In ISIS-V [DaZd86], at each transaction commit, a new version of the entire database is generated, defining a linear sequence of database states. All changes made during a transaction, schema or instance modifications, are stored in the new database version. Returning to previous definitions implies returning the whole database to a previous state.

5 Final remarks

This paper described a schema evolution mechanism considering the STAR data model. The proposed mechanism allows the definition of a schema and its modification in several ways, either adding or removing nodes/attributes/relationships between nodes. The mechanism is extremely flexible and capable of retaining the system's development history, based on versions of objects.

The implementation of the STAR database is being made using KRISYS [Matt91], a knowledge base management system that provides object-oriented concepts. The version layer is implemented. All operations here described have been completely specified and are being implemented as part of the data definition language.

Bibliography

[Atki89] M.Atkinson et al. *The object-oriented database system - manifesto.* Rapport Technique Altaïr 30-89, 21 août 1989.

[Bert91] E.Bertino. Object-oriented database management systems: concepts and issues. *Computer,* Los Alamitos-CA, v.24, n.4, p.33-47, Apr. 1991.

[DaZd86] J.W.Davison, S.B.Zdonik. A visual interface for a database with version management. *ACM Trans. on Office Information Systems,* v.4, n.3, July 1986.

[Deux91] O.Deux. The O_2 system. *Communications of ACM,* v.34, n.10, Oct. 1991.

[Gott88] K.Gottheil et al. The Cadlab Workstation CWS - an open, generic system for tool integration. In: F.J.Rammig(ed.). *IFIP Workshop on tool integration and design environments.* North-Holland, 1988.

[Groe91] K.Groening et al. From tool encapsulation to tool integration. In: F.J.Rammig, R.Waxman(eds.). *Eletronic Design Automation Frameworks.* Elsevier Science Publishers, 1991.

[KiCh88] W.Kim, H-T.Chou. Version of schema for object-oriented databases. In: *VLDB Conference,* 1988.

[Matt91] Mattos, N.M. *An approach to knowledge base management.* Springer-Verlag, 1991. (Lectures Notes in computer Science, 513).

[PeSt87] D.J.Penney and J.Stein. Class modification in the GemStone object-oriented DBMS. In: *Sigplan Notices,* December 1987.

[SkZd86] A.H.Skarra, S.B.Zdonik. The management of changing types in an object-oriented database. *Sigplan Notices,* v.21, n.11, 1986. (OOPSLA-86)

[Wagn92] F.R.Wagner et al. Design version management in the STAR framework. In: *3rd IFIP International Workshop on EDA Frameworks.* North-Holland, 1992.

[WaVi91] F.R.Wagner and A.H.Viegas de Lima. Design version management in the GARDEN framework. In: *28th. ACM/IEEE Design Automation Conference,* June 1991.

[WGF94] F.R.Wagner, L.G.Golendziner, M.R.Fornari. A tightly coupled approach to design and data management. In: *EURO-DAC 94.*

[Wilk88] W.Wilkes. Instance inheritance mechanisms for object-oriented databases. In: K.R. Dittrich(ed.). *Advances in Object-Oriented Database systems.* Springer-Verlag, 1988.

[Wolf88] P. van der Wolf et al. Data management for VLSI design: conceptual modeling, tool integration & user interface. In: F.J.Rammig(ed.). *IFIP Workshop on tool integration and design environments*. North-Holland, 1988.

[Zica91] R.Zicari. A framework for schema updates in an object-oriented database system. In: *International conference on Data Engineering*, 1991.

PART THREE

Data Types

6

Integration of a Domain Conceptual Model with the EDIF Information Model

Rachel Y.W. Lau and Hilary J. Kahn
Department of Computer Science
University of Manchester
Manchester M13 9PL
UK
email: rachel@cs.man.ac.uk, hilary@cs.man.ac.uk

1 Background

In the area of electronic design, EDIF (Electronic Design Interchange Format) is a well accepted standard for design exchange between different CAD systems. Information Modelling has been widely used by the EDIF community to describe the design information carried by an EDIF file. The EDIF Version 3 0 0 Information Model [1] represents formally the semantics of EDIF Version 3 0 0 and is the basis from which the EDIF Version 3 0 0 syntax [2] has been derived. In addition, the model provides the basis for the future extension of EDIF. Currently, work is in hand to extend EDIF into other domains such as PCB (Printed Circuit Board) and Test. The relevant EDIF Technical Sub-Committees were therefore asked to develop their domain conceptual models as a basis for integration into the core EDIF model in the future. Their domain conceptual models specify a set of requirements which should be met by future versions of EDIF. At present, the modelling activity is focused on creating an integrated model between the EDIF Version 3 0 0 Information Model and the PCB Conceptual Model [3] developed by the EDIF PCB Technical Sub-Committee.

This paper addresses several issues related to model integration, in particular on large-scale models. In the case of the core EDIF model and the PCB domain model, both models have hundreds of objects defined in their domain of interest. Hence, it is important to divide this large task into several phases. Figure 1 shows the process of integration between the PCB Conceptual Model and the EDIF Information Model.

Firstly, the PCB Conceptual Model undergoes a translation process so that it uses the same notation as in the core EDIF model. The model integration process is roughly divided into two phases. The first phase is to achieve the model integration between the core model and the PCB model at the conceptual level. The target of this phase is to establish possible mappings between PCB objects and EDIF objects. This phase requires active participation from the EDIF PCB Technical Sub-Committee (EDIF PCB TSC) and also the EDIF Technical Experts Group (EDIF TEG). The EDIF PCB TSC defines the scope of EDIF-PCB integration and also provides expert advice on the PCB domain. The EDIF TEG offers advice on EDIF issues which help to maintain conformity and consistency in any extension of EDIF. In many cases, there is more than one way to map the PCB objects into EDIF. Therefore experts from both committees need to decide which mapping should be used for the final integrated model. The second phase creates the fully integrated model based on the mappings identified in the first phase. In addition, it ensures coherence and consistency in the final integrated model.

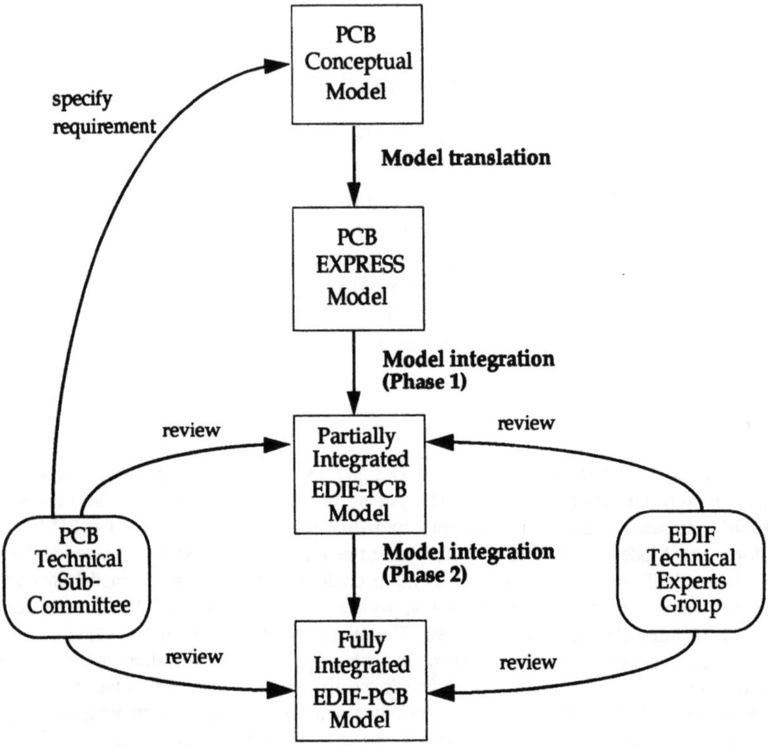

Figure 1: Integration process between the PCB Conceptual Model and the EDIF Information Model

2 Model translation

The first task of the integration is to make sure that the core EDIF model and the domain conceptual model use the same notation. The PCB Conceptual Model is used as the model integration example here. The PCB Conceptual Model is specified in both textual and graphical form. Each object defined in the PCB Conceptual Model is given a plain English textual description, together with a detailed list of attributes and the associated constraints. In contrast, the core EDIF model is specified formally in EXPRESS [4] - a modelling language which describes a domain in terms of schemas, entities and constraints. An EXPRESS schema defines a universe of discourse in which the defined objects have a related meaning and purpose. An EXPRESS entity defines an object of interest and is described by a number of attributes. An entity may have rules which constrain the value of individual attributes or a combination of attributes for every entity occurrence. In order to facilitate comparison between the formal core model and the semi-formal domain model, the PCB Conceptual Model is translated into EXPRESS so that it shares the same notation as in the EDIF model. Each PCB object is converted into an EXPRESS entity. The major advantage of this translation process is that the domain model will be formally defined and subsequently can be checked by existing EXPRESS checkers for correctness. Hence, the translation process makes sure that the domain model is self-consistent and unambiguous, before any integration process takes place.

The translation of PCB objects into EXPRESS is a fairly straightforward task. However, additional information is added to the PCB EXPRESS model. One useful aspect involved is to determine whether an attribute of an object indicates an ownership relationship or a reference relationship. This information is useful for deciding the top level entities in the PCB model - an essential phase in mapping between the PCB objects and EDIF objects. This is done by careful examination of attribute names and their supporting documentation. In the PCB Conceptual Model, if an attribute uses terms such as "has" or "owns", it may imply an ownership relationship. For example, a *pin_group* object has an attribute *has_pins* which is a set of *pins*. It implies a *pin_group* owns a set of *pins*. On the other hand, if an attribute uses terms such as "refers" or "is_made_of", it may imply a reference relationship. For example, a *physical_layer* object has an attribute *is_made_of* which is a *material*. The *is_made_of* attribute is documented as "A reference to the material of which the layer is made" therefore it is easy to deduce that a *physical_layer* references a *material*.

3 Model integration requirements

There are a number of requirements which should be met by the integrated model. These requirements are set in order to maintain the consistency of the core model and to protect the existing users of the core model. The major requirement is to maintain **upward compatibility** between the core model and the integrated model. The upward compatibility constraint can be stated as "The integrated model must be a superset of the core model".

In other words, the integrated model must be able to convey all the information in the core model and users of the core model are guaranteed that they will not be affected by any changes made on the core model after the integration. In order to achieve this, the EDIF community has introduced these additional requirements on the integrated model:

- All objects which are defined in the core model must remain in the integrated model. That includes schemas, entities, attributes and constraints in an EXPRESS model.
- All objects which are defined in the core model must not have their names changed in the integrated model.

4 Model integration (Phase 1)

The model integration process is divided into two phases. The target of the first phase is to create a partially integrated model between the core EDIF model and the domain EXPRESS model. The first phase integration process is to determine how domain-specific objects are related to the objects already defined in the core EDIF model.

Before attempting any integration, it is necessary to have a reasonable understanding of the models that need to be integrated. The core EDIF model is already well understood, but the PCB EXPRESS model is less known. In the process of becoming familiar with the domain EXPRESS model, general mappings between the domain model and the core model inevitably spring to mind. However, these general mappings may not be the final mapping since they may be based on an incomplete understanding of the models. The high-level general mappings can be established by identifying categories of design information in the PCB model and determining how they are related to the design information already defined in the core EDIF model. Several categories of design information have been identified as common between the two models. Their use and possible merging are discussed below:

- instantiatable object information - in the core EDIF model, this provides template definitions whose occurrences can be used to support the implementation part of cell views. Similarly, in the PCB model, several classes of sub-layout definition are defined to support the description of a reusable bare board. However, most EDIF templates are designed specially for the schematic domain and they should not be overloaded to describe templates used primarily in the PCB domain. Hence, PCB-specific templates are not merged with any existing EDIF templates.

- connectivity information - in the core EDIF model, connectivity is divided into logical connectivity and structural connectivity. The logical connectivity defines the entire connectivity for a given level of hierarchy and the structural connectivity describes the implementation of the logical connectivity. In the PCB model, three levels of connectivity are identified; these are the functional netlist, the component netlist and the physical netlist. The functional netlist describes the complete connectivity that is to be realised by an assembled board and it corresponds to the logical connectivity in EDIF. Similarly, the component netlist and the physical netlist can be classified as the structural connectivity. However, structural connectivity in EDIF Version 3 0 0 is domain-specific. Hence, the PCB component netlist and physical netlist are not merged with existing EDIF structural connectivity objects which are defined specifically for other domains, e.g. schematic.

- geometry information - the concept of *geometric_shape* in the PCB model corresponds closely to the concept of *geometry_element* in the core EDIF model. Hence, it is possible to map PCB *geometric_shape* directly to EDIF *geometry_element.*

After some general high-level mappings are identified, the integration process continues by finding mappings between entities in the two models. The general procedures of the EDIF-PCB entity integration process are explained below.

4.1 Identifying top level entities

The owner of each entity in the PCB EXPRESS model is identified. The purpose of this activity is to determine the top level entities in the PCB EXPRESS model. Top level entities are most likely to be the core or major concepts. They represent the objects of interest which are good starting points from which the mapping should begin.

However, finding all top level entities is not a simple task. It involves the manual examination of all attribute relationships defined in the PCB EXPRESS model. It uses the additional information given on attribute relationships in the model translation process (see section 2).

4.2 Identifying areas of overlap

Areas of commonality and overlap between the core EDIF model and the PCB EXPRESS model are located. This is needed because every entity (i.e. object) in the domain EXPRESS model must be either created as a new entity or merged with an existing EDIF entity in the integrated model. The criterion used is to merge with existing entities whenever possible. Some possible ways of entity integration are described below.

If a domain entity corresponds to an entity defined in the core EDIF model and it conveys less information than the existing EDIF entity, an obvious solution is to replace the domain entity by the existing EDIF entity in the integrated model. However, domain experts should be consulted on whether or not they actually want the extra facilities provided by this existing EDIF entity.

For example, the PCB *identification_stamp* entity records the identification information of a PCB object. It includes the dates when the object was created and last updated. On the other hand, in the core EDIF model, the identification information of an object is captured by the *written* entity which records the time when the object was created or last modified. In addition, the *written* entity may also include program identification, human or organisation identification, or location information which can help to trace the origin of an EDIF object. It is obvious that the PCB *identification_stamp* entity can be mapped onto the EDIF *written* entity. PCB domain experts are consulted in order to decide whether or not it is appropriate to use the existing EDIF entity. In this case, it is decided that a separate PCB *identification_stamp* is not necessary for the integrated EDIF-PCB model and the existing *written* entity should be used instead.

In cases where a domain entity corresponds to an entity defined in the core EDIF model and it includes additional information which is not covered by the existing EDIF entity, changes to the existing EDIF entity will be required. EDIF experts should be consulted to agree on the changes made to the EDIF entity. For example, the EDIF *font* entity specifies information about the appearance of a text font and it is usually a reference to an externally defined standard text font. On the other hand, the PCB *text_font* can be either a reference to an externally defined standard text font (i.e. *standard_text_font*) or a set of rules for drawing individual text characters directly (i.e. *non_standard_text_font*). Hence, the core EDIF model should be extended to describe the character-based text font required by the PCB domain. One possible integration is to extend the EDIF *font* entity to include attributes belonging to a character-based text font. However, this leads to overloading of the existing *font* concept which is undesirable. Consequently, the current solution is to add a new *character_based_text_font* entity in the integrated model which describes the PCB *non_standard_text_font*. The existing EDIF *font* and the new *character_based_text_font* are related by sharing a common supertype. The common supertype *text_font* contains attributes which apply to both subtypes, e.g. font height and font width.

4.3 Determining place-holders for top level entities

If a domain entity describes an object which is not currently modelled in the core EDIF model, the simple solution is to add this domain entity to the integrated model. For example, the PCB *package* entity describes the mechanical data (information such as body volume, number of pins) of a physical package. It is a new concept for the core EDIF model, so the *package* entity is added to the integrated EDIF-PCB model.

Any top level domain entities that are not merged with concepts in the EDIF core model need to be placed somewhere in the resulting model. The two most likely places that such entities will be placed are in the *edif* entity or in the *library* entity.

The *edif* entity is the only top level entity in the core EDIF model and amongst other things contains libraries. The *library* entity is used to group related information. For example, the TTL 7400 series of components can be grouped into one or more EDIF libraries based on attributes such as functionality and performance.

The main criterion used in deciding whether a top level PCB domain model entity is placed in *edif* or in *library* is: does the entity represent information that it would be useful to transfer independently of a complete design? To answer this question it is necessary to consult with domain experts.

5 Model integration (Phase 2)

The first phase integration process produces the partially integrated model detailing the changes required to the core EDIF model. The second phase of the integration process adds various EDIF features to the partially integrated model. The following are some of the important issues:

5.1 Adding internal/external distinction

In the core EDIF model, a *library* can be either internal or external to an EDIF file. An *external_library* represents a *library* which may be used within the file but is not explicitly defined with the file. That is, objects defined inside an *external_library* contain sufficient information to be used but they must not carry any detailed implementation information. On the other hand, objects defined inside an *internal_library* may contain the complete description. One of the reasons of making the internal/external distinction is that EDIF is designed for transferring design information between CAD systems. When design information is transferred across systems, it is desirable to transfer the same information only once. Hence, a transfer format needs a mechanism which supports references to previously transferred data. For example, a library of commonly used design information is transferred to the receiving system as an *internal_library*. In subsequent transfers, the sending system can refer to the previously transferred library at the receiving system as an *external_library*. This can lead to a considerable saving because duplicate transfer of design information is avoided.

The internal/external distinction in EDIF has made a considerable impact on the integration process. It is because in the PCB conceptual model, objects were defined without differentiating between interface information and implementation information. Thus, each attribute of a PCB entity needs to be examined carefully to determine whether it is part of the interface information or part of the implementation information. For example, in the PCB model, the *pin* entity has a *cross_section* attribute describing the cross-sectional shape of a pin, a *name* attribute describing the name of the pin and a *pin_length* attribute describing the length of the pin. Both the *cross_section* and the *pin_length* attributes are used to determine the physical shape of a pin; they are therefore classified as implementation information.

Figure 2 shows how the internal/external distinction is added to the *pin* entity. In the integrated

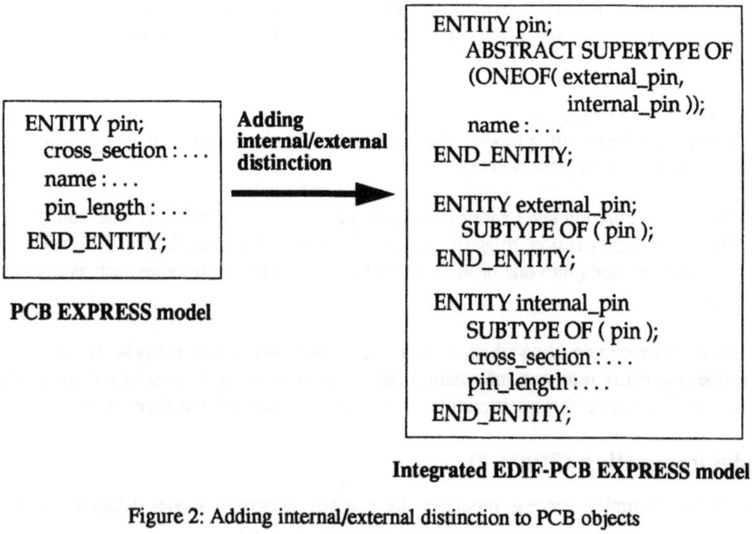

Figure 2: Adding internal/external distinction to PCB objects

EDIF-PCB model, a *pin* can be either an *internal_pin* or an *external_pin*. An *internal_pin* is defined within an *internal_library*; whereas an *external_pin* is defined within an *external_library*. The *name* attribute, which conveys the interface information, is placed in the *pin* entity. Both the *internal_pin* and the *external_pin* are subtypes of the *pin* entity and subsequently inherit the interface information. However, only *internal_pins* are allowed to contain implementation information, i.e. the *cross_section* and the *pin_length* attributes.

5.2 Adding units

In the PCB model, the units of measurement are not defined because it was felt that this level of detail was not required in a conceptual model. For example, the PCB model includes the concept of *distance* but does not specify whether this is measured in meters or inches. In EDIF, there is a *unit* mechanism which allows units to be defined in terms of SI units. Each measurement can then be associated with one of these user defined units. This mechanism is considered to be more than adequate by the PCB domain experts and so it is a simple task to merge any unit of measurement in the PCB domain model with the corresponding unit in the EDIF core model.

5.3 Adding name information and properties

EDIF has the concept of name information which is used to capture the object naming information in the CAD system. In addition, an EDIF object may have properties to convey CAD-specific information. PCB experts need to be consulted to determine which PCB objects have meaningful names or properties in CAD systems.

5.4 Maintaining consistency

When the EDIF-ised partially integrated model is added to the core EDIF model to create the final integrated EDIF-PCB model, it is important to maintain a consistent modelling style and naming mechanism in the fully integrated model. There are a number of issues which need to be considered:

- Names of entities and attributes should have a consistent style. For example, in the PCB model, the *package* entity has an attribute *has_pin_groups* which is a set of *pin_groups*. In the integrated model, the prefix of the attribute name "*has_*" is removed and *has_pin_groups* is renamed as *pin_groups*. It is felt that attribute prefixes like "*has_*" convey no additional information, so they are generally removed from the integrated model. This is consistent with the style of the existing EDIF Version 3 0 0 Information Model.

- There is a need to check that every entity added as a result of the integration process has an existence constraint defined either directly or indirectly. This is needed to ensure that no new entity is allowed to "float about" and the structure of EDIF is maintained.

- There is a need to check that the EDIF style of grouping mechanism is obeyed in any grouping constructs introduced in the integrated model. In EDIF, grouping is achieved by referencing the grouped objects. For example, an EDIF *signal_group* provides a grouping mechanism for *signals* and contains a list of references to other *signals* or *signal_groups*. However, in the PCB model, a *pin_group* "contains" a collection of *pins*. Consequently, PCB *pin_group* does not conform with EDIF style of grouping mechanism. Under these circumstances, the problem needs to be discussed between domain experts and EDIF experts so that they can decide whether or not a violation of the general mechanism should be permitted in the integrated model. In this case, it is agreed that the EDIF style of grouping mechanism should be used for modelling *pin_group*.

6 Problems encountered during model integration

During the integration process for PCB, new requirements have been added by the PCB domain experts. This has substantially increased the scope of the integration. These new areas included modelling of assembly drawing, technology rules and MCM (Multi Chip Modules). Hence, careful consideration must be given to ensure that the integrated model can be easily extended to cover these new areas.

Sometimes, it is difficult to decide which mapping should be chosen for a PCB object when there is more than one viable way to integrate it with EDIF. For example, in the PCB model, a *component* is an instantiation of a *part*. One possible mapping is to model *part* as a new class of instantiatable object in EDIF. Another possibility is to integrate it with the existing EDIF instantiation mechanism. The core EDIF model has the concept of instantiation: an *instance* is an instantiation of a *cluster* in a *cell*. Initially it seemed natural to adopt the second approach which maps the PCB *component* to the EDIF *instance* and the PCB *part* as a new type of view in a *cluster*. It was only much later in the integration process that it was decided this mapping is not appropriate because a PCB *part* is not a hierarchical object whereas an EDIF view contains instances of other cells to build a design hierarchy.

One of the model integration requirements listed in section 3 is that no object defined in the core EDIF model is allowed to have its name changed in the integrated model. This restriction also caused some discussion in the integration process. For example, the EDIF *font* entity currently describes an externally defined text font and it corresponds to the PCB *standard_text_font* entity. Ideally, the *font* entity should be renamed as *standard_text_font* because the new name describes exactly what it is. Also it is less confusing when the PCB *character_based_text_font* is introduced. However, name changing is considered to be an issue which affects other EDIF domains such as schematic. As a result, names of existing EDIF objects have not been changed in the current integrated model.

7 Conclusions

Having to merge quite distinct models in one go can be difficult. By dividing the integration process into two phases, the task of merging becomes manageable. Also, it is easier for both EDIF experts and domain experts to identify and review the changes to the core EDIF model.

Based on the experience derived from the EDIF-PCB integration, it can be concluded that the first phase of the model integration requires much more time and effort than the second phase. This is because during the first phase, various ways of mapping between objects need to be identified and the best possible integration has to be chosen. Sometimes, in order to evaluate trade-offs between various mappings, it is necessary to present several partially integrated models so that it is easier for the reviewers to gain an overall picture of different ways of integration.

The integration of a domain conceptual model with the EDIF Information Model requires a good understanding of both models. This understanding cannot be derived from the EXPRESS models alone since much of the necessary information is not contained in an EXPRESS model but is to be found in accompanying documentation. For example, a *part* in the conceptual model is identified as an instantiatable object because the description of *component* in the accompanying documentation states that a *component* is an instantiation of a *part*. This information cannot be directly derived from the attributes of *part* and *component* or any relationship between them. The fact that much of the information needed to merge models is not in a formal form that can be readily processed by a computer means that complete automation of the model merging process is not possible. More experience in merging models manually needs to be obtained to identify common areas in which there is potential for automation.

Finally, it is important to note that the integration procedures outlined in this paper should be taken as guidelines only. Since it is not possible to predict what format or form any other domain conceptual models (e.g. Test domain, VHDL, CFI) will take, it is difficult at this stage to come up with a general principle which can be applied in all possible cases.

8 Bibliography

1. Electronic Industries Association. *Electronic Design Interchange Format Version 3 0 0*, EIA-618, Volume 1, Dec 1993.

2. Electronic Industries Association. *Electronic Design Interchange Format Version 3 0 0*, EIA-618, Volume 2, Dec 1993.

3. EDIF PCB Technical Sub-Committee. *Conceptual Model of a PCB Version 12*, Nov 1993.

4. International Standard Organization STEP ISO/TC 184/SC4/WG5. *EXPRESS Language Reference Manual*, N14 edition, April 1991.

7

Generating Abstract Datatypes with Remote Access Capabilities

Martin Schuetze, Bernd Schuermann, Joachim Altmeyer
University of Kaiserslautern, Germany

Abstract:

We propose a software system to specify data management components of CAD tools on the modelling level. The implementation of the resulting abstract datatype will be done by a set of generators. Functions to access data stored outside the abstract datatype are also generated. The information models of the ADT and the external data may differ; the transformation between both can be defined in a mapping table and will be handled by the generated code.

1. Introduction

An important aspect of ECAD framework technology is to provide efficient and easy-to-use access to design data for the different tools. While design tools can be supplied with a limited amount of data (for the current design task) in a single transfer, e.g. via files, design management or analysis tools potentially need access to all design data stored by the framework. Usually, this access is provided through a library of access functions to external data sources (e.g. databases). The programmer of management tools or design tools has to implement code for the data structures the tool works on as well as code to translate between the tool's data structures and the information model of the accessed library.

We investigated the structure of the design tools in the framework of our PLAYOUT VLSI Design System /Zim89, SAP94/. All tools are structured as it can be seen in figure 1. The data needed by a tool are stored in an abstract datatype (ADT). The graphics component, if part of the tool, displays data stored in the ADT. The control component determines the internal control flow, while the user interface lets the designer interact with the tool. The investigation showed that all shaded components are very similar from tool to tool. For example, the ADTs of different tools manage different data models, but the way objects are accessed is the same in all ADTs. Because of their supporting nature for the design algorithms we call these components *framework components*. The algorithms are highly individual. But while the development of algorithms is a creative task, a lot of time is needed to implement and test the other components which is lost for the algorithm development.

Our approach to overcome these problems is to specify the framework components on the modelling level and let the implementation be done by a generator. Because this work is most advanced in the field of data modelling, we will consider only this aspect in the following sections. For the data management components, we can specify a mapping between local data structures and external data structures in addition to the information model. A set of generators creates the corresponding ADT, including methods to access data stored outside the tool.

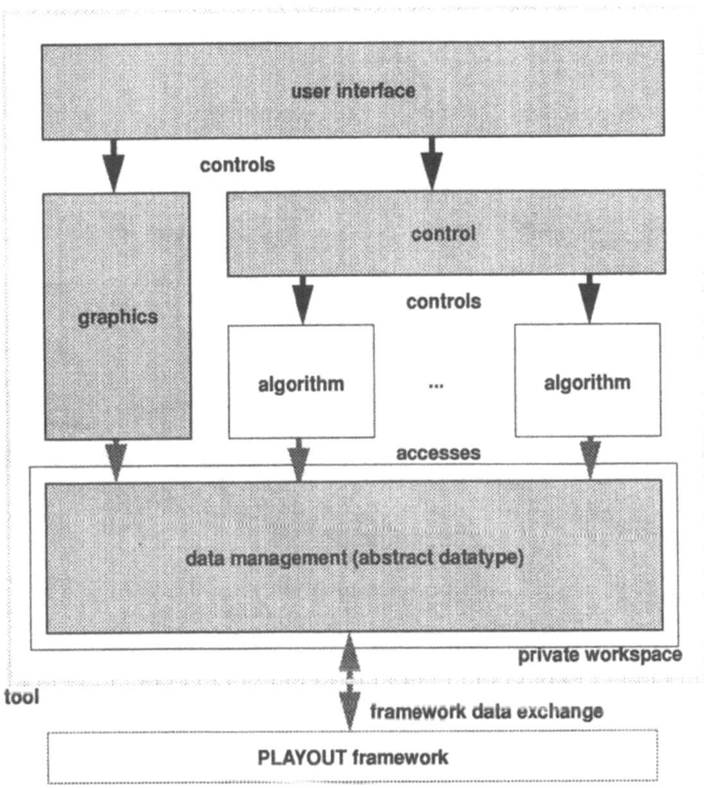

Figure 1 Architecture of a Tool

2. Data Management within the PLAYOUT Framework

The PLAYOUT framework, like many others, consists of several tools grouped around a database which holds the design data (see figure 2). We will refer to the database as the ***archive workspace*** or the ***server***. The design tools (the algorithms) themselves do not work on the database directly. They access their data in the private workspace (the ***client***) via a programming interface of the generated ADT. The private workspace holds the local tool data and can be seen as a cache of the server data. The data are transferred via files or via socket communication between the server and the clients. A data exchange between private workspaces is also possible. In this case, one workspace acts as a server to the other private workspace. The data for the clients are generally accessed in a check-out-check-in manner. All transaction, consistency, and version control is done within the server and trans-

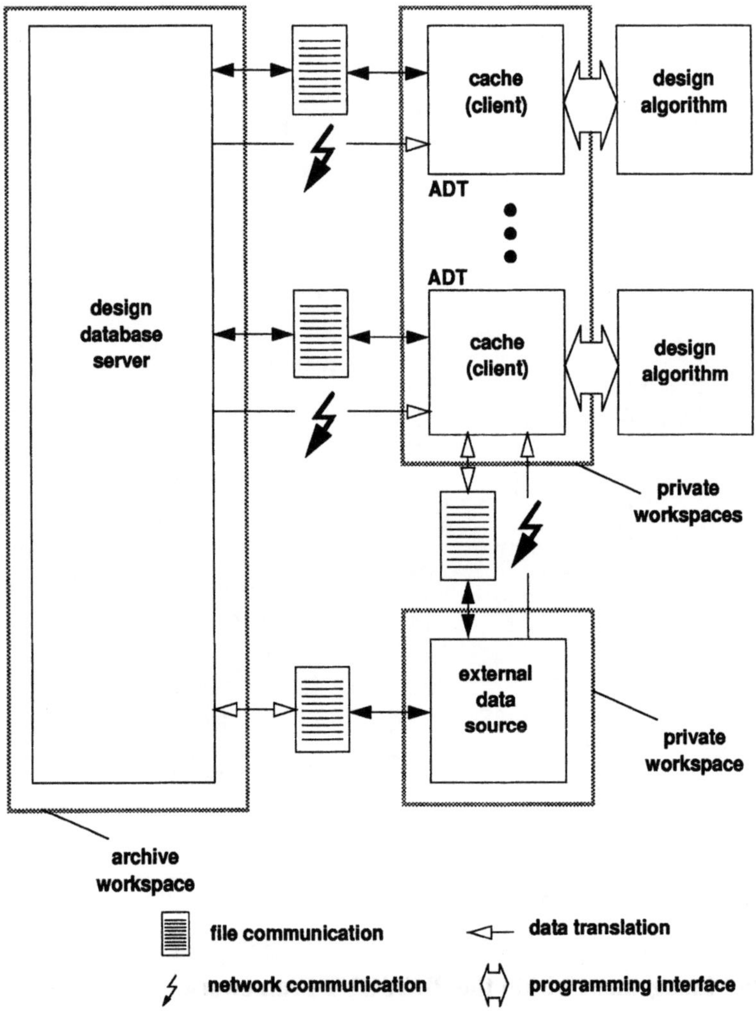

Figure 2: Data Management in the PLAYOUT framework

parent to the tools. The data exchange between client and server is managed by the workspaces and not visible by the tools. The data management described here acts as a infrastructure which is exploited by the generated code for the framework components.

3. MOOSE

The modelling of framework components as well as the code generation is done within the MOOSE system (**M**odel-Based **O**bject-**O**riented **S**oftware Generation **E**nvironment). This section describes the development of appropriate models as well as the structure of MOOSE and the generated code.

3.1 Models

All framework components have to be described by models which are input to the generators. Our current implementation of MOOSE allows the graphical definition of information models with extended entity-relationship-diagrams (EER). The development process of the information models is divided into two steps. In the first step, a ***base model*** describing the whole application domain of the framework (in this case ECAD) has to be found. This idea is shared by many approaches to model the VLSI design process /HRS92, BrK92, CFI92/. However, this base model is often not very well suited for the implementation of tools. It usually covers more aspects of the framework than a single tool needs. In addition to that, most tools need local data structures not contained in the base model. We therefore modify the base model in a second step with tool specific ***characteristics*** to achieve the ***application model*** which describes all data needed for the tool.

3.2 MOOSE architecture

The architecture of MOOSE can be seen in figure 3. The user enters the base model and the characteristics of an application with the graphical editor. The editor checks that only syntactically correct information models are entered. The information models are kept in an ADT (which has been generated by MOOSE, too). Several generators are provided to generate different types of output: The ***C generator*** produces an ADT of the application model in the programming language C. The ***C++ generator*** produces a C++-ADT, respectively. The ***network communication generator*** allows access to external data sources in conjunction with the C++ generator. It is described in more detail in section 4. The ***Versant generator*** creates code for the commercial OODBMS Versant. The ***hypertext generator*** creates FrameMaker hypertext documents for the tool programmer with an implementation description of the information models. The ***file communication generator*** generates code to save and restore data of the ADT to/from disk. These files can either be used to checkpoint applications or to exchange data between different ADTs. More generators for different framework components (graphics, user interface) are under development.

3.3 Generated code

All ADT code generators offer nearly the same functionality for the tool programmer. They create the following methods/functions for operations on the objects:

- ***Object creation***
- ***Object deletion***
  If an object is deleted which is still referenced by other objects, the reference will destroyed. Undefined references are avoided.
- ***Attribute access functions***
  Attribute values can be set or read.
- ***Relation access function***

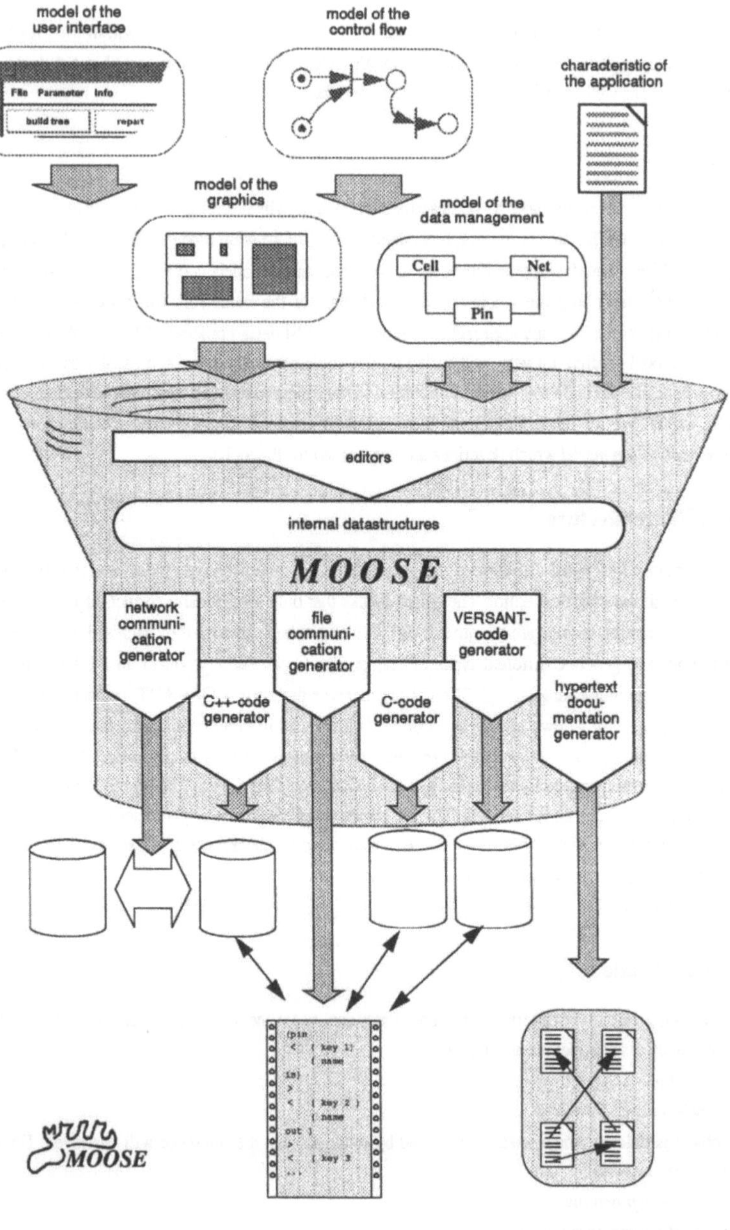

Figure 3: MOOSE System

Relations between object instances can be established or deleted. The objects participating in a relation can be retrieved from the object holding the relation.

- *Checkpointing*

  Methods are provided to save the state of the workspace to disk or to restore data from disk.
- *Meta informations*

  Objects can be queried about their type and about the relations and attributes provided by this type. Access methods can be sent dynamically at runtime to these objects by a string-to-method-mapping.
- *Remote access capabilities*

  The ADT can remotely access data stored outside the private workspace. At the moment, only read access to these data sources is allowed. Write access will be added in the future. Differences in the local and the remote information model can be described in mapping tables and are also handled by the generated code.
- *Other operations*

  There are a number of other operations defined on objects, types, and workspaces such as pretty printing and workspace initialization.

All generated classes can be further extended by the user. Methods can be added to the ADT in additional files. They will not be overwritten if the ADT is generated again.

Our approach to the generation of workspaces has some similarities with the generation of database schemata, particularly for object-oriented databases /ODM93/. However, our framework architecture provides one single, central design database. Most tools aren't and shouldn't be implemented on top of a database directly. They work on local data structures only and don't need database support as multi user support, transactions, or atomicity during execution. The MOOSE system provides a workspace functionality and 'comfort' for the tool programmer as some databases do, but without the overhead usually introduced through the database concepts. As stated above, these database concepts are provided by the server. However, if a single tool needs direct database support, this can be achieved with the Versant code generator.

4. Remote Access Capabilities

Workspaces created by the C++ generator can exchange data via two different file formats: PLIF and SLANG. SLANG (Simple Data Description Language) provides a general structure to describe objects, attribute values, and relation values. The exact file format is defined during the generation process of the ADT and can be used to store and restore the workspace contents efficiently, as well as to exchange data between generated ADTs. PLIF (Playout Interchange Format) is a defined file format (similar to EDIF) for the exchange of data between all design tools. Most of our design tools access data through PLIF because it is an easy and efficient way to transfer all data needed for one design step. All data needed for one of these tools can be determined a priori by the server and can be retrieved in one single query which makes the transfer process easy and efficient.

However, for some tools we can make no assumptions about the data they need at runtime. Design management and analysis tools potentially need access to all design data stored in the server. It is more efficient if they retrieve their data dynamically during runtime from the server. This is often achieved through libraries allowing to access, for example, databases by using SQL statements, leading to the following problems:

- The information models of the ADT and the external data source do not match. The programmer of CAD tools is forced to translate between the different data structures.
- The programming language and the retrieval language may differ.
- The library is not updated automatically in case that the information model for either the server or the cli-

ent are changed.

Our network communication generator partially overcomes these problems by generating the necessary methods for remote access along with the other ADT methods. One important goal is to keep the access as transparent as possible. Ideally, the tool programmer 'sees' only the tool's application model and doesn't need to know about the server's information model.

4.1 Architecture

The generated access methods are part of a client server architecture as it can be seen in figure 4. An access to

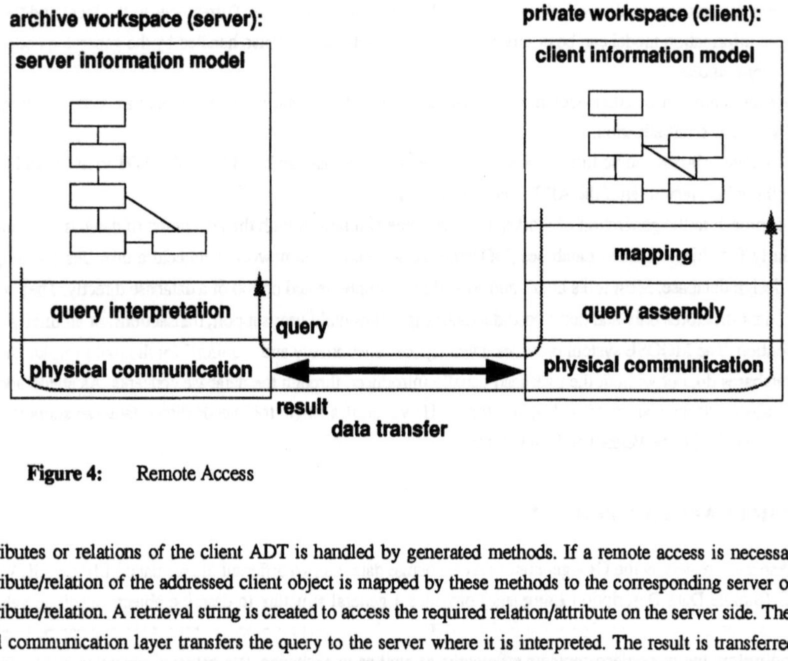

Figure 4: Remote Access

attributes or relations of the client ADT is handled by generated methods. If a remote access is necessary, the attribute/relation of the addressed client object is mapped by these methods to the corresponding server object's attribute/relation. A retrieval string is created to access the required relation/attribute on the server side. The physical communication layer transfers the query to the server where it is interpreted. The result is transferred back into the client and, after conversion to the client's data structures, delivered as the result of the originally called client access method. At the same time, the accessed data are cached for later use.

In order to establish the remote access service in the private workspace, we need:

- A *communication link* between client and server

  We have chosen the UNIX socket mechanism. Other techniques could also be used, e.g. RPC libraries.
- A *query language*

  Objects must be accessed in the server. This is achieved through a query language. Strings over this language must be composed at the client side and interpreted at the server side. We use the query language of our design database.
- A *key* to identify objects uniquely on both client and server side

  The key must be defined on all client objects and on all server objects accessible by the client. It may be a simple number or a pointer as well as complex string. We have chosen a hierarchical key composed with respect to the object (partOf-) hierarchy of the design objects.

4.2 Mapping between two information models

As stated above, our generator is capable of generating code which has a limited capability to map between different information models on client and server side. It is obvious that not all differences between two models can be handled by this generator. In this subsection we will explain which types of mapping can occur within our framework and which of them can be handled by our generator.

Because the mapping can not be derived from the client information model, it has to be defined in a mapping table. This table may contain an entry for every attribute and every relation of all object types. There are three ways to specify the mapping between client and server models: the client objects's attribute/relation name is mapped to a name on the server side (***name-to-name-mapping***). Or the client's name is mapped to a complex query for the server (***name-to-query-mapping***). A third possibility is to specify C++-Code which is put into the generated code by the generator (***complex mapping***). This is the most 'manual' mapping but it allows to deal with nearly arbitrarily complex mapping problems.

Attribute mapping

The following cases can occur while accessing attributes:

- The object type in the server's information model might have more attributes than the corresponding object type in the client's information model.
  We simply ignore these additional informations.
- The object type in the server's information model might have less attributes than the corresponding object type in the client's information model.
  We solve this problem by assigning initial values to attributes not available on the server side.
- Attributes may have different names in different information models.
  We map the client object's attribute name to the server object's attribute name (name-to-name mapping).
- Attributes are stored in other server objects.
  This case may appear if the server information model supports more object types than the client, and the client combines these server objects types to one client object type. If the object holding the attribute can be accessed unambiguously from the current client object, a simple query expression can be used to map the attribute (name-to-query mapping). Otherwise, the mapping must be done with a C++-statement (complex mapping).

Relation mapping

Basically, the relation mapping works exactly as the attribute mapping. However, there are two more problems which can arise:

- Relations may have different cardinalities.
  If the cardinality of the relation on server side is a subset of the cardinality of the relation on the client side the mapping can be handled with a name-to-name mapping. If there are more objects in the server object's relation than in the client object's relation and a subset can be selected at every time, either a query expression or a C++-statement can be used for the mapping. All other cases need manual interference after the generation of the access functions.
- Relations may only exist on either client or server side.
  If a missing relation in the server can be substituted by the evaluation of a set of other relations, the access function can again be generated by using name-to-query mapping. If such a query cannot be found for a certain relation, it is not possible to generate remote access code for it.

Object mapping

A mapping between groups of objects (split or merge of object types) is currently not supported by the generator. These problems must either be solved by mapping each attribute or relation individually as stated above (which is sufficient in many cases). Or it must be solved on the modelling level, e.g. by creating the appropriate object types in the application model.

4.3 Current Implementation

Our current implementation of the network communication generator allows the definition of application models and mapping tables with MOOSE as stated above. The access functions are generated as declared in the mapping table. We currently allow only remote read access until a new multi-user synchronization technique is fully implemented on the server side. However, write operations are already supported by the transportation layer and our query language.

An application model may contain parts which access the server as well as parts managed only locally. This can be determined on a per-relation or per-attribute basis. Once the ADT is generated, the remote access is nearly invisible to the tool programmer. However, if the programmer wants to, he may manipulate the caching scheme as well as he may inquire if an object is created locally or remotely accessed. If the server or client information model is changed, all depending ADTs can be consistently updated by modifying the mapping table and re-generating and recompiling the ADT code. Although the definition of complex mappings requires some manual work, it is still helpful because it allows the generation of the access method body and automatically inserts the complex mapping.

5. Related Work

OODBMS grant transparent access to objects in distributed repositories. At least databases conformant to the ODMG /ODM93/ standards allow the generation of ADTs from an information model. Although our description and generation techniques are similar, we can not use this type of database directly. We believe that design data management in a framework is the task of the central database and not a tool issue. Tools within the PLAYOUT framework are therefore created without direct database support at runtime.

As far as we know, OODBMS do not support a transparent access to data stored in different databases with different information models. Chomicki and Litwin /ChL94/ describe a possibility of object mapping between different databases by an extension to the OSQL language. They distinguish 16 different types of mapping. Many of these mapping types are similar to the mappings we propose. However, they concentrate on multi-database mappings only, therefore, their approach can not be used directly.

Newer techniques for the management of distributed objects (e.g. OMG /OMG92.1, OMG92.2/, DCE /RKF92/) allow to access objects distributed over a network of heterogenous computers. However, they don't support the programmer with an automated mapping between different organizations of data. It is usually neither possible do define and generate an ADT, nor are the problems of different information models addressed. However, we intend to use these techniques for the communication layer in the next version of our generator.

Software Engineering Environments allow the definition of information models, object hierarchies, and dynamic aspects of an application. However, the description techniques used are intended to document the specification of components and are often too general to generate ready-to-use application code. We do not know of any SE environment supporting our concepts of base models, application models, and framework components.

6. Conclusion

We showed that it is possible and feasible to specify and generate ADTs with remote access capabilities. Through the principle of generation, we can always guarantee a consistent set of ADTs derived from the same information base model. The techniques described in this paper are currently used in different applications of our CAD framework. The access to other workspaces is nearly transparent to the tool programmer. Because of the simplicity of the transportation layer and the query language, other servers than databases (e.g. other tools) could be used. This could support cooperative strategies between private workspaces.

MOOSE and its generators can be used in many applications. Once the base model is defined, a great amount of work is saved with every new tool introduced or with every change to the data model. Measurements within our framework show that the data management components make up 25% to 40% of the tool code. Generating this code saves implementation time and makes MOOSE useful for 'normal' application development as well as for rapid prototyping or the development of experimental systems. The generation of remote access methods is not intended as a replacement for database technology but as an alternative way to load the private workspace with data; the tools still process data locally.

In the future, we will add new capabilities as automatic object key management, remote write access, automatic propagation of data changes, and more support for the mapping between information models as soon as we develop the proper relation semantics.

And finally, the integration of models and generators for the other framework components (like the user interface or the control component) will enable us to generate a complete tool skeleton with MOOSE, leaving the creative part for the tool programmer: the development and implementation of design algorithms.

References

/BrK92/ M. Brielmann, E. Kupitz, "Representing the Hardware Design Process by a Common Data Scheme", Proc. Int. European Design Automation Conference, 1992

/CFI92/ "Design Representation Electrical Connectivity Information Model and Programming Interface", CFI Pilot Release Document, CFI-92-P-6, 1992

/ChL94/ J. Chomicki, W. Litwin, "Declarative Definition of Object-Oriented Multidatabase Mappings", in M. T. Özsu, U. Dayal, P. Valduriez, "Distributed object management", Morgan Kaufmann Publishers, 1994

/HRS92/ C. Huebel, D. Ruland, E. Siepmann, "On Modeling Integrated Design Environments", Proc. Int. European Design Automation Conference, 1992

/ODM93/ T. Atwood, J. Duhl, G. Ferran, M. Loomis, D. Wade, "The Object Database Standard: ODMG-93", edited by R.G.G. Cattell, Morgan Kaufmann Publishers, 1993

/OMG92.1/ R. M. Soley (ed.), "Object Management Architecture Guide", Rev. 2.0, OMG TC Document 92.11.1, John Wiley & Sons, 1992

/OMG92.2/ "The Common Object Request Broker: Architecture and Specification", Rev. 1.1, OMG Document 91.12.1, John Wiley & Sons, 1992

/RKF92/ W. Rosenberry, D. Kenney, G. Fisher, "Understanding DCE", O'Reilly & Associates, 1992

/SAP94/ B. Schürmann, J. Altmeyer, H. Pahle, M. Schütze, "A Framework for the VLSI Design System PLAYOUT", GI/ITG/GME-Fachtagung "Rechnergestützter Entwurf und Architektur mikroelektronischer Systeme", Oberwiesenthal, 1994 (in German)

/Zim89/ G. Zimmermann, "PLAYOUT - A Hierarchical Design System", Information Processing 89, G.X. Ritter (ed.), Elsevier Science Publishers B.V. (North Holland), IFIP, 1989

8

Extending View Technology for Complex Integration Tasks[1]

Matthew C. Jones and Elke A. Rundensteiner
Software Systems Research Laboratory
University of Michigan, Ann Arbor, MI 48109-2122
e-mail: mjones@eecs.umich.edu, rundenst@eecs.umich.edu

Abstract

In this paper, we present extensions to the MultiView object-oriented database view system that improve its ability to integrate electronic computer-aided design (ECAD) tools. Using an object-preserving algebra as the view definition language, the MultiView system supports data transformations to suit the needs of individual tools. However, an object-preserving algebra alone is not powerful enough to express recursive transformations such as transitive path derivations and the flattening of hierarchical structures. In order to provide these capabilities, we extend the view definition language with operators that permit complex transformations of the data. To achieve performance essential for ECAD tools, we introduce the mediator class as a general strategy for the materialization of these complex views. We compare the bounds on query and update performance of our set-based mediator class for the symmetric transitive closure with a more traditional object-based materialization strategy.

1 Introduction

Object-Oriented Databases (OODBs) are typically chosen to implement ECAD databases because they support the modelling of complex data [1,3,14,16]. Although integrated ECAD tools benefit from these database services, such as data integration and access control, these tools often employ different internal representations of design data. A schematic editor, for example, represents graphical entities hierarchically in the design schematic, including information about position and orientation. On the other hand, an analysis tool might represent the design as nodes and edges in a flattened graph. Even with current integration standards [4,6] the burden of transforming design data to and from the central database format falls upon the tool developer. This manual and often ad-hoc transformation process is labor-intensive and error prone. An integration system should provide a means for automating these data transformations. Additionally, if tools are to operate cooperatively and concurrently, tools are needed to simplify the communication of data changes between tools.

Recently, OODB *view systems* have been presented as a means for providing automatic reorganization of central data that has been customized according to a high-level restructuring request (i.e., a query). In this model, the tools in the system are *view users*, using central data that has been automatically organized by the view system. Our OODB view system, MultiView, [14] provides a *view definition language* (VDL) capable of restructuring data for each tool in the system. As such, it automates the transformation of data between the central format and the tool specific formats. In addition, the system provides the services necessary to maintain consistency between the central and derived data [11]. MultiView assures the correctness of data transformations and reduces undesired coupling between integrated tools. As a consequence, it increases the productivity of tool developers and integrators. However, the MultiView system is currently not capable of performing complex data transformations, such as the computation of closure, the traversal of paths, or the flattening of hierarchical graph structures.In this paper we present new operators for the VDL of the MultiView system. These operators permit it to fill the role of a central data store in a tool integration environment supporting complex data transformations. To provide the performance essential for ECAD applications, we introduce the *mediator class*, our abstraction for storing and maintaining the results of the transformation associated with a view operator. We address the design and performance of the mediator class associated with the computation of the symmetric transitive closure (STC) derivation.

In Section 2, we present view technology as the basis for tool integration. In Section 3, we discuss the MultiView system. In Section 4 we present the TC/STC operators and describe characteristics of the base and derived data. In Section 5 we explain the implementation of both an object-based and our mediator-based strategy for materializing the STC operation. We compare the performance of the mediator with the more general object-based materialization strategy in Section 6. We briefly describe other types of mediators in Section 7, and cover related work and conclusions in Sections 8 and 9.

1. This work was supported in part by NSF NYI #IRI-9457609, NSF RIA #IRI-9309076, Intel, and Digital Equipment Corporation.

2 Applying Object-Oriented View Technology to Tool Integration

Object-oriented view mechanisms [1, 14] have been proposed as a means to provide customized data reorganization allowing access to centrally stored and managed data (*base data*) that is automatically restructured or *derived* into a preferred format. The view is considered *updatable* if changes to the derived data produce corresponding changes in the base data. Recently there has been considerable research into OODB view technology. However, with the exception of the O2 system [3] in alpha-test, commercial OODBS are not currently providing support for views.

Figure 1 illustrates the use of a view system as the basis for tool integration. A *data server* manages centrally stored data. A view system produces customized views in the form of *derived data*, generated by the view system for each tool (*view user*) in the environment. If the view system is distributed, each of the views is implemented by a client called the *view maintainer* that provides the view user access to the derived data and communicates changes between the data server and the view.

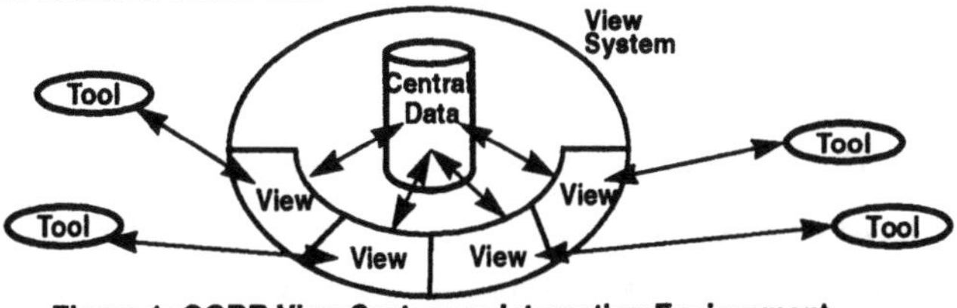

Figure 1: OODB View System as Integration Environment.

The view system may operate as a vehicle for the check-in and check-out of tool data - providing reorganization when tools read and write data. It may also act as a run-time integration system by propagating changes back to the base data (and subsequently to other active views) as changes are made to derived data.

The implementation of such a system requires solutions to *many* problems in data modelling, database views, and distributed systems. We identify key requirements for a view-based integration environment below, and describe where they fit within the scope of our work.

1. The tool integrator must be able to specify how data is to be reorganized. This *view definition* is usually specified using a view definition language (VDL). The VDL must be expressive enough to define complex views appropriate for ECAD applications. Our work extends the MultiView VDL accordingly.
2. The system must be able to create the view as defined by the tool integrator and provide the view user with access to the derived data. The access can be via a *materialized view* (implemented as actual view objects derived from base objects in the central database), or non-materialized view (implemented as an interface to base objects). Because performance is crucial in many ECAD applications, and because it is possible to express complex data transformations with the VDL, we present a solution that supports the materialization of derived data by the view maintainer.
3. If the view user makes changes to derived data, the system must propagate these changes to the base data (*view update*). The problem of *view consistency*, i.e., propagation of base data updates to views, only occurs when views are materialized. In such a case, the central data server must notify the view maintainer to make the appropriate changes to the derived data. We assume the existence of a notification mechanism, and accomplish view consistency by notifying the view maintainer through an update request which has been appropriately transformed for the associated view.
4. The tool integrator must be able to specify when and if updates are propagated to and from a view. This is accomplished using *update contracts*. Update contracts will be discussed in a future paper.

In this paper, we present an *extended query language* for defining complex view transformations, as well the development of a key abstraction for the materialization and maintenance of these complex views.

3 MultiView: The Object View System

At the University of Michigan, we have an on-going research project funded by NSF and Intel for the development of an object-oriented view support system, called MultiView [14]. MultiView has been successfully used to integrate ECAD tools in the domain of behavioral synthesis [15]. It is particularly suited to integration tasks such as hiding unneeded details from a tool, simplifying application specific processing by augmenting a view with customized functions, and precompiling information which is frequently needed by the tool.

For example, a simple placement tool may only be interested in adjacencies between components rather than the complete circuit netlist as stored in the central database. A view defined in MultiView can abstract away the component pins and explicit pin and net connections, and provide a derived attribute that models the adjacency relationship. In this case, MultiView is hiding the unnecessary details of the pins, simplifying processing by encapsulating the adjacency attribute, and precompiling the adjacency information needed by the placement tool.

The MultiView system currently employs an *object-preserving* algebra as a query language for view definitions. Among its unique features, MultiView: (1) provides for full property inheritance for both base and virtual classes, (2) automates integration of virtual classes into a global schema, and (3) supports generation of complete view schemata rather than just virtual classes. The first prototype has been fully functional for over a year. It currently runs on a Sun SPARCstation and has been implemented on top of the Gemstone OODB [10].

To model the more complex restructuring employed in ECAD tools, such as flattening hierarchical graphs and deriving transitive relationships, we are now proposing extensions to MultiView. We have designed extensions to the query language to support more complex, recursive queries. In addition, we are designing extensions to the set-based data model to include sequences, graphs, and paths. Because of these added complexities we also must re-examine the implications to query processing and the view update and consistency problems.

In this paper, we enable the derivation of views optimized for ECAD applications by adding complex, object-creating operators. This optimization is twofold: (1) transformations on ECAD data are compact and easy to specify, reducing effort during integration, and (2) efficient access to the transformed data, providing the performance that is critical in ECAD applications.

4 Extending the VDL for Closure Computation

4.1 Characteristics of the Base Data

We now present the language extensions by which these complex views can be specified. We also characterize the kinds of derived views possible with these new operators.

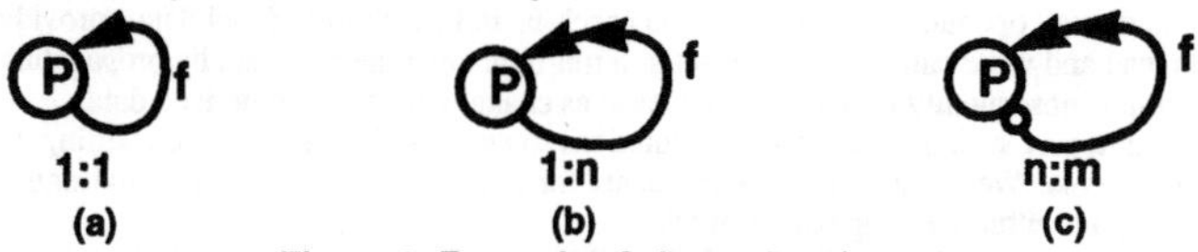

Figure 2: Recursive Schema Graphs.

Consider views derived from the base schemata illustrated in Figure 2. These schemata contain a single class `P`, and a relation **F** (modeled by the instance variable `f`) that relates two instances in `P`. We define how the instance variable `f` models the relationship **F** formally as:

$$\text{for } p_1, p_2 \text{ in } \mathtt{P}\text{:} \quad p_2 \text{ in } p_1.\mathtt{f} \quad \text{iff} \quad p_1 \, \mathbf{F} \, p_2 \tag{1}$$

The cardinality of this relation determines the properties of instance graphs possible with this schema. For example, an instance of the schema in Figure 2.a would result in a set of sequences of instances of `P`. Because these instance graphs have known properties, we are able to select our materialization strategy based upon this knowledge. Independently of whether or not the relationship **F** is transitive, the view user may wish to derive new attributes based upon the transitive closure of the attribute `f`. We consider two possible transitive derivations over the attribute `f`: the *symmetric* and *asymmetric transitive closure* known as STC_f and TC_f respectively.

4.2 Characteristics of the Derived Data

The derivation of TC/STC creates a derived attribute with more general semantics than the base attribute. For example, if the attribute `f` represents the relation **F** (Fanout) between two instances in `P` (`Parts`), then the attribute `c` derived from the TC on `f` represents the more general semantics of **C** (Fanout-Cone). The transitive closure on the specialized relation **F** is used to derive the more general relation **C**. In this example, Fanout is a specialization of the Fanout-Cone relation. We derive the more general TC_f relation as follows:

$$\text{for } p_1, p_2 \text{ in } \mathtt{P}\text{: } p_1 \, \mathbf{F} \, p_2 \text{ implies } p_1 \, \mathbf{C} \, p_2 \tag{2}$$

$$\text{for } p_1, p_2, p_3 \text{ in } \mathtt{P}\text{: } p_1 \, \mathbf{C} \, p_2 \text{ and } p_2 \, \mathbf{C} \, p_3 \text{ implies } p_1 \, \mathbf{C} \, p_3 \tag{3}$$

Figure 3 illustrates an example of a schema, a possible object graph, and the derived TC_f instance graph. Both equations (2) and (3) are applied to the base instance graph to produce the derived instance graph. The application of equation (2) represents attributes *directly derived* from the base data. Equation (3) produces the attributes indirectly, or *transitively derived*, from base data. Note that the instance variables shown in the schema are represented by edges in the instance graph. The instances of the class `P` are nodes in both the base and the derived graph. This means that changes to the instance variables `f` in the base objects correspond to edge deletions and insertions in the base and derived instance graphs.

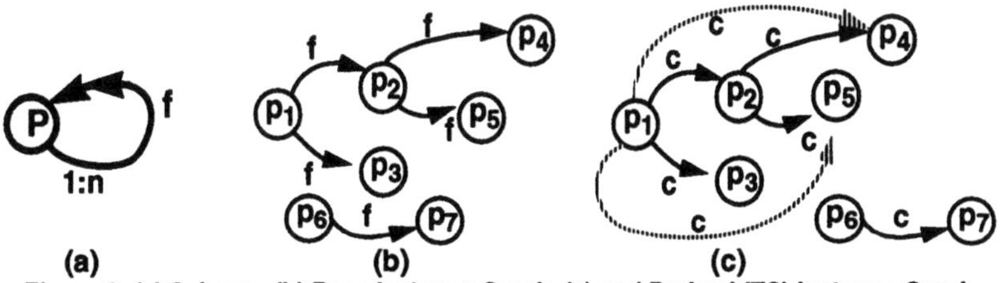

Figure 3: (a) Schema, (b) Base Instance Graph, (c) and Derived (TC) Instance Graph.

Note that a similar generalization occurs with the derivation of the STC. Here we derive a new relation **B** (In-Block) that relates all parts in a common combinational circuit block. To derive the STC, the relation **B** is directly derived from **F** by applying the derivation in (2), then it is *symmetrically derived* by applying:

$$\text{for } p_1, p_2 \text{ in } \mathtt{P}\text{: } p_1 \mathbf{B}\, p_2 \text{ implies } p_2 \mathbf{B}\, p_1. \tag{4}$$

and then transitively derived using formula (3). The resulting derived STC, a set of fully connected graphs for the base instance graph of Figure 3b, is shown in Figure 4. The derivation for an instance graph of N objects where $N \equiv np$ divides the graph into n disjoint partitions connected by the derived attributes. The average size of each of these partitions is p.[1] If the base instance graph is connected by the base relation, then we have $p=N$ and $n=1$.

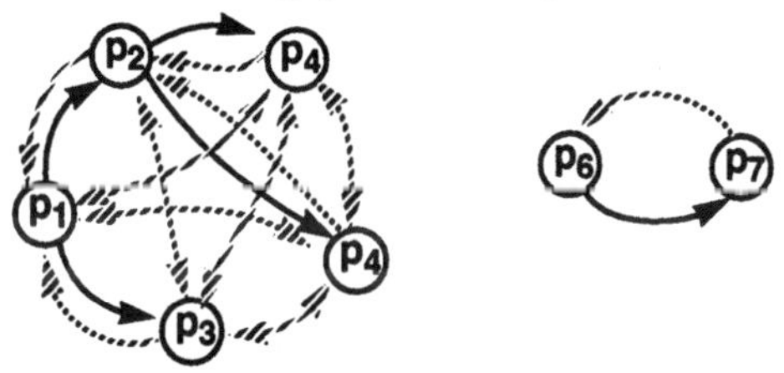

Figure 4: STC$_f$ Instance Graph Derived from Figure 3.b.

4.3 Updates to/from Derived Relationships

The attributes such as those derived by equations (2), (3), and (4) are generalizations of the base attributes from which they are derived. The newly derived attributes represent a loss of information, and thus are called *lossy.* This has important implications for the derived relations and their use in a view system. The most important implication is that it is impossible for a view maintainer to incrementally update derived information that has "lost" information which contributes to the data derivation. We nullify the effects of the lossy derivation by recording enough information in the materialized view to distinguish between directly and indirectly derived attributes. This ensures that updates can be propagated, and that the impact on the view of base updates can be autonomously computed by the view maintainer. In general, it is not always possible to propagate updates made by the view user to indirectly derived attributes. This problem is addressed in more detail in Section 6.4.

5 Efficiently Materializing Closure Views with the Mediator Class

To achieve performance critical in ECAD, we employ view materialization. This materialization is invisible to the user of the view, however, the choice of representation for the materialized view can have a dramatic effect on the efficiency of queries and updates to the derived data.

Using the more traditional view materialization approach as utilized by relational systems, the materialization of a new relation is accomplished by creating a new object (tuple) representing a pairwise relationship between two objects. The result is a new class $\mathtt{T_{P,B}}$ whose instances contain two attributes p_a and p_b whose domain is $\mathtt{P}$. An instance is placed into the extent of class $\mathtt{T_{P,B}}$ for each derived relationship. For this object-based materialization representing the derivation of the new relation **B**, we define the extent of $\mathtt{T_{P,B}}$ as follows:

$$\text{for } p_a, p_b \text{ in } \mathtt{P}\text{: object}(a{=}p_a, b{=}p_b) \text{ in } \mathtt{T_{P,B}} \text{ iff } p_a \mathbf{B}\, p_b. \tag{5}$$

1. For our analysis in the remainder of the paper, we assume that all subgraphs in the view have the same size p.

Both the size and time required to materialize this newly derived class are $O(n \times p^2)$. This bound comes from the fact that for each instance subgraph in the derived instance graph there are potentially p^2 newly derived relations. Recall that number of these subgraphs in the derived instance graph is n. The appeal of this object-based approach is its generality and simplicity. It can be used for any kind of derived relation. The data manager can manipulate this form of materialized view easily because it is a primitive form. We have found that one of its significant drawbacks is its inefficiency (both in space and time).

We thus introduce the *mediator class* as an alternative means to materialize the newly derived relation in the view. A mediator class is an optimized representation of the derived relation that accurately models the relation, and is able to propagate updates from the base data to the derived data efficiently. The mediator class is selected based upon known properties of the base data, and known properties of the derived relationship. The creation and materialization of a mediator class requires the following steps:

1. Identification of the appropriate mediator class for the current view of the base and derived data.
2. Materialization of the newly derived relation in mediator class objects.
3. An interface to the mediator to make it "look and feel" like the expected relationship semantics. These include access functions to query derived instance variables and methods to update them.

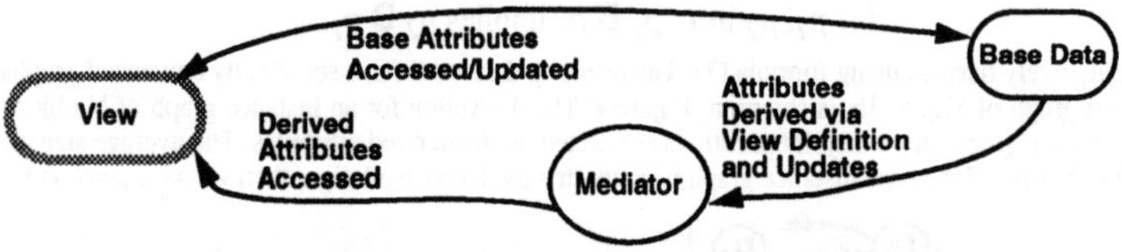

Figure 5: Mediator Class for Complex View Definition.

We assume that the view system knows enough about the characteristics of the base data to select a mediator class for the available query operators. Steps 2 and 3 are the focus of the sections that follow. In our mediator-based solution to materialization, we extend the derived class with a derived attribute that provides access to the mediator. The view system provides query and update operations on the derived attribute so that it has the appearance of a genuine attribute.

6 Implementation of Mediator Classes for Complex Views

This section discusses the implementation of a mediator class for the materialization of the STC. We analyze the performance of the implementation, and compare it with a simple object-based materialization approach.

6.1 Symmetric Transitive Closure

A user may wish to define a view comprised of the symmetric, transitive closure operation of an attribute. In an ECAD application, this type of transformation is typically used to combine a graph into a single entity, i.e., to gather segments in a schematic into a single net. We refer to the original schemata as illustrated in Figures 2.b and 2.c, and to the relations and attributes presented in Section 4.2. Our mediator implementation for the STC_f is independent of the cardinality of the attribute **f**, so we present the general solution here. Optimizations specific to a certain cardinality are discussed when they are significant.

We consider the user request to extend the class **P** with a derived attribute **b**, resulting in a virtual class **P'**.

```
P' = refine [ b = STCf(self) ] for (P).
```

Because of the nature of a symmetric, transitive relationship, the view system materializes the derived attribute **b** via a set-based mediator. The derived attribute **b** can be materialized using a set because of similarities between the properties of sets and the definition of symmetric, transitive closure. As expressed in the query, the new attribute **b** will provide access to the derived STC_f(**P**). **Same-set** membership in this set can then be used to model the appropriate transitive and symmetric relationships because of the following:

$$\text{Symmetry: } p_1 \text{ in } \textbf{same-set} \text{ as } p_2 \text{ implies } p_2 \text{ in } \textbf{same-set} \text{ as } p_1 \quad (6)$$

$$\text{Transitivity: } p_1 \text{ in } \textbf{same-set} \text{ as } p_2 \text{ and } p_2 \text{ in } \textbf{same-set} \text{ as } p_3 \text{ implies } p_1 \text{ in } \textbf{same-set} \text{ as } p_3 \quad (7)$$

Because same-set membership is a transitive and symmetric relationship, we define STC_f(**P**) to be of a set type, and materialize the resulting sets as instances of the mediator class. We also provide an interface to the mediator class to query and propagate updates to and from the view efficiently. Each object in **P'** has associated with it an instance of the mediator class. This mediator class is made available through the attribute **b** that is added with the view definition. Figure 6 shows the base instances with the derived attribute **b** associating each instances with the mediator class instances used to represent the symmetric transitive closure relation. The mediator accurately represents the derived relationship by satisfying the following constraint:

$$\text{for } p_1, p_2 \text{ in P: } p_1 \text{ in } p_2.\mathtt{b} \text{ iff } p_1 \text{ in } STC_f(p_2) \quad (8)$$

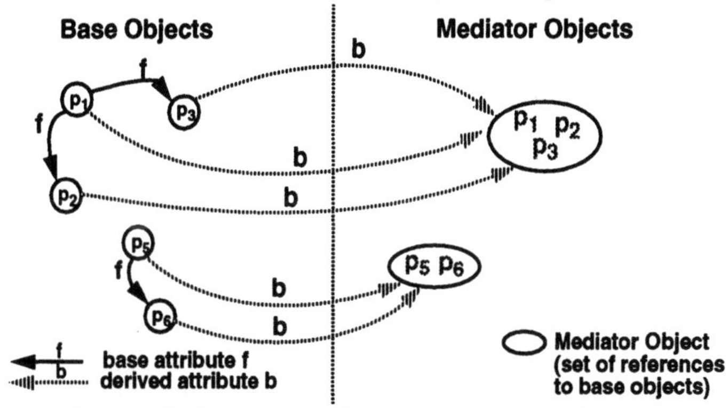

Figure 6: Mediators and the Derived Attribute b for STC$_f$(P).

6.2 Creation of the Derived Attribute and Mediator Class

The traditional object-based materialization approach derives objects to represent the STC. These objects are materialized in the class $T_{P,B}$ using the following operations:

```
create_class TP,B subclassOf Object with (a:P, b:P);
insert (select a:p1.self, b:p2.self from p1, p2 in P
        where p1 in STCf(p2)) into TP,B;
```

We create our mediator class with the following queries:

```
create_class M subclassOf STCMediator;
insert (select distinct the STCf(p) from p in P) into M;
```

In both cases, a new attribute is created for the class P' which provides the interface to the derived and materialized STC relation. For the mediator case, the creation of the b attribute for the class P' is as follows:

```
P' = refine [ b = (select m from M where p in m)] for (p in P);
```

This defines the derived attribute to provide access to the mediator object that maintains the STC relation. For the object-based materialization the new attribute provides the interface to the query and update operations needed to access the STC materialized in $T_{P,B}$.

6.3 Querying

We compare here the relative costs of performing queries on the traditional object-based materialization of the STC and our mediator-based materialization.

6.3.1 Membership Test in the STC$_f$(P)

We next show how the two representations answer the query: `Is y in the STCf(x)?`

Object-Based Representation: This tests for existence of an object that reflects the relationship between x and y.

```
return not empty(select * from TP,B where a=x and b=y);
```

Because the size of the extent of $T_{P,B}$ is $O(n \times p^2)$, this operation can be performed in time $O(\log(n \times p^2))$. This assumes the existence of proper indices to facilitate the search. The assumption of these indices impacts the estimated cost of insertions and updates, as well.

Mediator-Based Representation: The mediator determines if y is in the STC$_f$(x) with the $O(1)$ comparison of the reference attributes to the STC sets to see if they refer to the same set object.

```
return (x.s==y.s);
```

If both x and y have the same mediator set, then they are by definition in the STC of each other.

6.3.2 Retrieval of the STC$_f$(P) Relationship

We now discuss how the two representations answer the query: `What is the STCf(x)?`

Object-Based Representation: In this case, we need to locate all objects in $T_{P,B}$ that contain **x**.

```
return select b from T_P,B where a=x;
```

Although the size of the extent of $T_{P,B}$ is $O(n \times p^2)$, and this operation takes $O(\log(n \times p^2))$ time, the actual performance may be bounded by the size of the STC which is returned. Since the number of objects which represents the STC is $O(p^2)$, this may become the bound on the performance of this query if p is sufficiently large. Because of the representation inherent in the object-based solution, it is not possible to return a reference to the set of objects in better than $O(p^2)$

Mediator-Based Representation: The STC is computed by returning the mediator set associated with x. Because the set is already grouped into a single object, a reference to the set can be returned in constant time. However, because the size of this set is $O(p)$, that is the bound on the performance of the query.

6.4 View Updates to Complex Derived Attributes

In general, the derived attribute **b** has ambiguous update semantics, as shown in Figure 7. This figure shows objects p_1 through p_5 in the base data related by the attribute **f**. The $STC_f(P)$ derives the fully-connected instance graph with the derived attribute **b**. Although neither materialization strategy uses an explicit graph to represent the derived data, it is useful for this example. The ambiguity in the view update occurs when the view user attempts to change the value of the derived attribute indicated with the α. If this reference to p_4 is removed from the attribute p_1.**b**, it is not possible to determine how to propagate this change to the base data. A single change to an indirectly derived attribute has many interpretations involving changes to base attributes. It might be possible for the view system to assess the minimal impact update necessary to propagate, but such a computation is expensive and of limited use. Merely changing the derived attribute value without propagating the change to the base is not an option, because that means the view no longer represents the $STC_f(P)$. After such a change it would no longer be possible to use the STC mediator, because the constraint on which the contents of the mediator is based is violated. For this reason, updates to the indirectly derived attributes are not permitted for the STC.

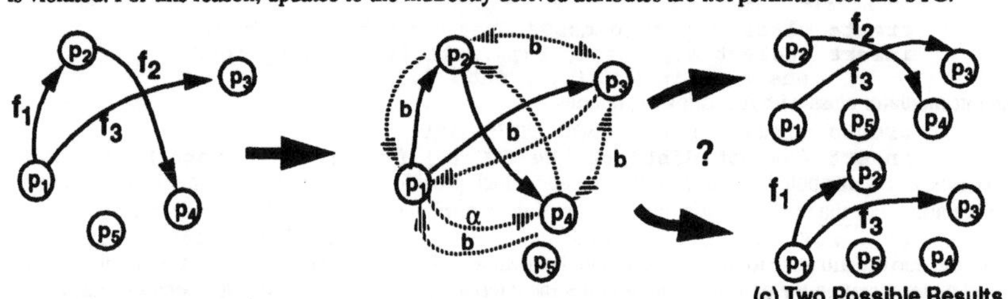

(a) base instance graph (b) Derived $STC_f(P)$ (c) Two Possible Results from Removing α

Figure 7: Ambiguous Update to Derived Attribute (deleting α).

6.5 View Consistency

If several tools use the base data simultaneously, then there must be a means to propagate updates performed on the base data to the views. We describe two updates which have direct impact on the derived view.

6.5.1 Reference Removal

We consider the effect of removing a reference to object y from the base attribute **f** of object x. This corresponds to an *edge deletion* in the base instance graph. The removal of the edge in the base instance graph may partition the graph, resulting in a corresponding partitioning of the sets materialized in the STC mediator.

Because the object-based approach is lossy, the view maintainer cannot *autonomously* determine the effect of such an update. As a consequence, the object-based approach requires that the central data manager communicates the resulting partitions to the view maintainer. The mediator class distinguishes between base and derived attributes, and consequently has enough information to autonomously compute the effect of the update on the view. The base data manager just reports attribute changes to the mediator. The mediator employs the same algorithm that would be used by the base data manager to determine the repartitioning of the base instance graph.

The partitioning algorithm builds two spanning trees of the base instance graph starting from both x and y. If the spanning trees are disjoint, then the attribute change partitions the graph, and the results of the partitioning must be propagated to the derived data. This check requires $O(p)$ time to complete. If the attribute does not partition the base instance graph, no further action is required (the STC remains unchanged).

Object-Based Representation: The view maintainer receives update information from the base data manager in the form of two sets S_j and S_k, representing the partition of the instance subgraph produced by the removal of the reference to *y*. The following method updates the class $T_{P,B}$ so that it reflects the removal of the reference to *x*.

```
foreach v in Sj
  foreach w in Sk
      remove object(a=v,b=w), object(a=w,b=v) from TP,B
```

The nested outer loops require $O(p^2)$ time, and the inner operation requires $O(\log(n \times p^2))$ time, for a total complexity of $O(\log(n \times p^2) \times p^2)$.

Mediator-Based Representation: After determining if the change to *x*.f partitions the base graph (and as a consequence computing the partitions S_j and S_k), the mediator updates its derived data with the following algorithm:

```
delete STCMediator(x.b);             // remove old set
new STCMediator(Sj);                 // create new sets
new STCMediator(Sk);
foreach v in Sj
  v.b = STCMediator(Sj);
foreach w in Sk
  w.b = STCMediator(Sk);
```

This algorithm associates each object with the new set that represents the STC of the object. It requires $O(p)$ time. In the case that the cardinality of the base attribute f is 1:n, we have a trivial partitioning of the base into the single element *y*, and the set consisting of x.b-y.

6.5.2 Reference Insertion

We consider the effect of inserting a reference to object *y* into the base attribute f of object *x*. This corresponds to an *edge insertion* in the base instance graph. The insertion of the edge in the base instance graph may result in the merger of two subgraphs in the STC derivation. Unlike edge deletion, however, the data manager takes no special actions to notify the maintainer of the STC view. It merely notifies the view maintainer about the addition to the attribute.

Object-Based Representation: The object-based representation receives update information in the form of an addition to the attribute value for the object *x*. The following method updates the class $T_{P,B}$ so that it reflects the new value for attribute f of object *x*. It accomplishes the merger of the two subgraphs by creating an object associating every element of the subgraph in which *x* is a member, to every element of the subgraph in which *y* occurs.

```
foreach w in { select b from TP,B where a=x }
  foreach z in { select b from TP,B where a=y }
      insert object(a=w,b=z) into TP,B
      insert object(a=z,b=w) into TP,B
```

The outer loops have complexity $O(p2)$, and the inner operation requires $O(\log(n \times p2))$ time, for a total complexity of $O(\log(n \times p2) \times p2)$. The insertion time in the inner loop could be constant, but that would be at the expense of increased cost to other operations. The logarithmic insertion time is consistent with the overhead of maintaining an efficient structure which supports fast queries and deletions.

Mediator-Based Representation: The following algorithm updates the mediator:

```
newset = new STCMediator(x.b union y.b);
delete STCMediator(x.b);
delete STCMediator(y.b);
foreach v in (newset)        // link elements to new mediator
  v.b = newset;
```

This solution simply associates each affected object with the newly merged subgraph which represents the STC of the object. It requires $O(p)$ for the algorithm.

The table in Figure 8 shows clearly that the mediator-based materialization strategy achieves advantages in both time and space performance over the traditional relational materialization approach. The mediator also demonstrates efficient membership and retrieval performance, a property essential for ECAD analysis tools that do many more reads than updates.

7 Other Mediators

We have discussed the implementation of the mediator for the STC operator in the Section 6. In this section, we briefly discuss the TC mediators. Mediators for the derived TC differ from the set-based STC mediator, since the symmetry modeled by the set-based mediator is not appropriate when materializing the TC relationship.

Metric	Mediator-Based	Object-Based
Space Required	O(n x p)	O(n x p^2)
Creation Time	O(n x p)	O(n x p^2)
Membership Test	O(1)	O(log (n x p^2))
STC Retrieval	O(1)	O(log (n x p^2))
Edge Deletion	O(p)	O(log (n x p^2) x p^2)
Edge Insertion	O(p)	O(log (n x p^2) x p^2)

Figure 8: Comparison of Mediator-Based and Object-Based Materialization Strategies

7.1 TC Mediator for 1:1

We refer to the schema illustrated in Figure 2.a, and consider the view definition:

```
P' = refine [ c = TCf(self) ] for (P).
```

If the cardinality of the attribute **f** is 1:1, we can materialize and maintain the TC_f with a mediator class containing ordered sets (or sequences). For each element *e* in the sequence we define a sequence number seqnum(e), that identifies the position of each element *e* in the ordered set. We then model the TC_f implicitly using the same-set test along with the seqnum(e) associated with each element *e*. We state the definition of the TC_f in terms of operations on the mediator as:

$$\text{for } p_1,p_2 \text{ in P: } p_2 \text{ in } TC_f(p_1) \text{ iff } p_2 \text{ in same-set as } p_1 \text{ and seqnum}(p_2) > \text{seqnum}(p_1) \quad (9)$$

An object-based materialization of the TC_f performs similarly to the object-based materialization of the STC_f so we will not discuss the performance of the object-based materialization in this section.

The mediator provides a membership test (see Section 6.3.1) on the materialization of the TC_f by comparing references to the associated ordered set and comparing sequence numbers. This can be accomplished in $O(1)$ time. The $TC_f(p_n)$ can be retrieved for an object p_n in constant time, provided that the mediator is capable of returning a reference to a subsequence in the ordered set, otherwise the operation requires $O(p)$ time to build the sequence to return.

Updates on the base data can be efficiently propagated to the mediator. If an edge in the base instance graph is removed, this results in a partitioning of the ordered set materialized in the mediator. The edge itself provides the point in the sequence where the cut must be made. The result is two sequences representing the two new partitions in the base instance graph. This operation can be accomplished in $O(p)$ time. If an edge in the base instance graph is added, this results in the merger of two base instance graphs. The mediator merges the corresponding ordered sets by concatenating them. The operation requires $O(p)$ time.

We materialize the TC for instance graphs containing loops by using a set representation similar to the STC mediator objects. Because all objects in a loop are transitively related to each other, the set models this relationship accurately. Updates that break loops eliminate the associated set-based mediator and create a corresponding ordered-set mediator. Similarly, the mediator handles updates which introduce loops (i.e., joining the end of a sequence to the beginning) by creating the corresponding set-mediator and removing the ordered set mediator. These operations require $O(m)$ time.

7.2 TC_f for Other Cardinalities

The precise materialization strategy for TC_f of other cardinalities of **f** is still an open question. We are currently investigating strategies for materialization which balance performance for both queries and updates. We can improve the performance of queries by utilizing topological ordering in much the same way that we employ sequence numbers for the 1:1 case.

8 Related Work

The Gandalf system [8] was one of the earliest environments to use views as the basis for tool integration. This system used "display-only" views to provide the data for various tools in a software development environment. Updatable views were proposed in this work, but not implemented. In the FICOM system [2], a data manager is presented that supports incremental consistency maintenance between different levels of abstraction in a CAD data manager. However, the representations are limited to those already implemented in the data manager. Furthermore, the FICOM system does not support materialization within the integrated tools.

In [9], the authors present the super-key class as a mechanism for enabling updates on materialized paths in object databases. Although the model they present is general, no implementation is discussed.

9 Conclusions and Future Work

In this paper, we have discussed extensions to the view definition language of MultiView to provide complex transformations as required by ECAD tools. We have proposed the addition of a mediator class to provide for the efficient materialization of certain classes of complex views. In addition, we presented the development of efficient materialization strategies which include support for updatable views. These contributions provide essential foundational technology for an integration support tool in an ECAD framework.

We plan to continue the development of update semantics for these complex structures, as well as classify the types of transformations which are appropriate for the integration of ECAD tools. This includes the evaluation of materialization strategies for the transitive closure operation, and other path transformations. We are also developing the use of our mediator class as a super-key class for more general path materialization.

Acknowledgments. Matthew would like to thank Karem Sakallah for financial support while he was "finding this topic". Many thanks go to Harumi Kuno and Tricia Jones for reviewing an earlier draft of this work.

10 References

[1] S. Abiteboul and A. Bonner, "Objects and Views," in *Proc. of the ACM SIGMOD 91*, 1991, pp. 238-247.

[2] R. Armstrong and J. Allen, "FICOM: A Framework for Incremental Consistency Maintenance in Multi-Representation, Structural VLSI Databases," in *Proc. IEEE International Conference on Computer-Aided Design (ICCAD)*, 1992, pp. 336-343.

[3] O2 Technology. *O2 Views User Manual*, version 1 edition, December 1993.

[4] CFI-DR-TSC, "Design Representation Electrical Connectivity Information Model and Programming Interface," *CFI Publication*, Version 1.0.0, 1992

[5] CFI-ITC-TSC, "Inter-Tool Communication Architecture," *CFI Publication*, Version 1.0.0, 1992

[6] L. Claesen, R. Severyns, P. Six, W. D. Rammelaere, H. D. Man, J. Cockx, P. Reynaert, and G. Shrooten, "Open Framework of Interactive and Communicating CAD Tools," in *Tool Integration and Design Environments*, F. J. Rammig, ed., North Holland, 1987.

[7] M. Jones, and E. A. Rundensteiner, "Mediator Classes for the Efficient Materialization and Update of Complex Views", Electrical Engineering and Computer Science Dept., University of Michigan, Ann Arbor, Tech. Rep., In Preparation.

[8] D. Garlan, "Views for Tools in Integrated Environments," in *Advanced Programming Environments*, Springer-Verlag, 1986, pp. 314-343.

[9] S. Konomi, T. Furukawa, and Y. Kambayashi, "Super-Key Classes for Updating Materialized Derived Classes in Object Bases," in *Proc. DOOD Conference*, Dec. 1993.

[10] H. A. Kuno and E. A. Rundensteiner, "Developing an Object-Oriented View Management System," *IBM CASCON*, Oct. 1993, pp. 548 - 562

[11] H. A. Kuno and E. A. Rundensteiner, "Materialized Object-Oriented Views in MultiView," in *Proc. Fifth International Workshop on Research Issues on Data Engineering: Distributed Object Management (RIDE-DOM '95)*, March 1995.

[12] J. Miller, K. Gröning, G. Schulz, and C. White, "The Object-Oriented Integration Methodology of the CADlab Work Station Design Environment," in *Proc. IEEE/ACM Design Automation Conf. (DAC)*, 1989.

[13] Y. G. Ra, H. Kuno, and E. A. Rundensteiner, "A Flexible Object-Oriented Database Model and Implementation for Capacity-Augmenting Views", Electrical Engineering and Computer Science Dept., University of Michigan, Ann Arbor, Tech. Rep. CSE-TR-215-94, May 1994.

[14] E. A. Rundensteiner, "MultiView: A Methodology for Multiple Views in OODBs," in *Proc. of International Conference on Very Large Data Bases (VLDB)*, 1992, pp. 187-198.

[15] E. A. Rundensteiner, "Design Tool Integration Using Object-Oriented Database Views," in *Proc. IEEE International Conference on Computer-Aided Design (ICCAD)*, 1993, pp. 104-107.

[16] M. H. Scholl, C. Laasch, and M. Tresch, "Updatable Views in Object-Oriented Databases," in *Proc. DOOD Conference*, Germany, Dec. 1991.

PART FOUR

Federated Systems

9

Framework and Multi-Formalism: the ASAR Project

P. Asar*
Correspondence should be sent to Michel Israël
LaMI Université d'Evry, Bld. des Coquibus, 91025 Evry Cedex
email: israel@univ-evry.fr

Abstract

The main concern of the research project ASAR is to build a multi-formalism framework oriented towards Architectural Synthesis. This paper describes the CENTAUR system, used to build this framework, and the common intermediate data-flow format GC, that will be the common denominator of the different formalisms and tools available in that framework.

Keywords: Framework Technologies, Exchange Formats, Multiple CAD tools integration, Intermediate Data Flow Graph.

1 Introduction

This paper describes a research project—named ASAR[1]—grouping together six french research teams. A main concern of this project is Architectural Synthesis Framework and user-interface management.

Most of these teams have been already involved in architectural synthesis for a long time, using distinct input formalisms and approaches. While independently developed, these approaches are complementary, and could be used during different stages of a design, or for designing different parts of a system.

The goal of the ASAR project is to build a generic architectural synthesis framework providing:

- the possibility of experimenting various approaches, corresponding to different target architectures and design description languages,
- multiple associated tools for transformations and synthesis.

This multi-formalism framework is currently under construction. It is built upon the CENTAUR [4] system, already used by three teams as a common generic software environment for user-interfaces, formalisms and tools management.

2 ASAR structure

The overall structure of the ASAR framework is sketched in figure 1. The ASAR framework provides a set of languages and tools oriented towards particular target architectures or domains.

SIGNAL [12] and LUSTRE [8] are data-flow oriented synchronous languages especially designed for real-time applications. They can be used to specify hardware/software systems at different levels: system level

*P. Asar is a generic name for the ASAR project members, namely, P. Aubry (IRISA), M. Auguin (I3S), M. Belhadj (IRISA), J. Benzakki (LaMI), T. Bouguerba (LaMI), C. Carrière (I3S), G. Durrieu (CERT-Onera), Th. Gautier (IRISA-INRIA), M. Israël (LaMI), P. Le Guernic (IRISA-INRIA), M. Lemaître (CERT-Onera), E. Martin (LASTI), P. Quinton (IRISA-CNRS), L. Rideau (INRIA-Sophia), F. Rousseau (LaMI), O. Sentieys (LASTI).

[1] ASAR stands for "Atelier d'accueil générique pour la Synthèse ARchitecturale".

and behavioral level for instance. They provide SIGNAL/LUSTRE to C compilers and proof checking tools. The SIGNAL environment, which includes a mixed graphic/textual program entry, provides a multi-level VHDL generator prototype (behavioral—that allows the use of synthesis tools such as Synopsys, RTL, gate level).

TRANSE [7] is an interactive transformational tool based on the LUSTRE language: a circuit is obtained through successive semantic preserving rewritings of LUSTRE programs, until good operational properties are obtained.

ALPHA [13] is a functional language based on recurrence equations, for the design of parallel hardware accelerators; it allows the expression of regular algorithms and/or architectures and the synthesis of regular architectures by stepwise refinement.

GAUT [14] and OSYS [9] are high level synthesis tools from behavioral VHDL descriptions. GAUT is dedicated to signal processing designs; OSYS is a general purpose prototype tool.

A more complete presentation of these tools can be found in [2].

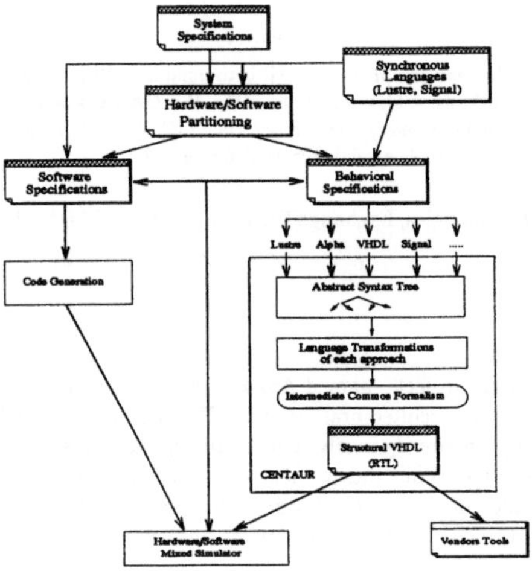

Figure 1: Overall structure of ASAR

Three components of ASAR are of main importance:

Centaur The CENTAUR generic environment is the first stone on which we build this common framework, allowing to manipulate different syntaxes (abstract, concrete) and user interfaces in an uniform way.

An Internal Common Formalism (ICF) In order to federate our different approaches based on distinct user-formalisms, a sound semantic basis is needed, into which each formalism can be expressed. In other words, there is an underlying hidden common semantic to our multiple user-formalisms, to be conveyed by this ICF. This common formalism, a kind of common denominator, need not be human-readable, and could serve as a common exchange format between different tools.

As our different approaches are all synchronous, a synchronous data-flow semantic seems to be the most appropriate. It is the reason why the *GC* formalism is under investigation for the ICF. GC [15] is a common format for synchronous data-flow languages, developed in the french *projet SYNCHRONE*.

VHDL The VHDL language, which is now a standard [1], cannot be ignored by any architecture oriented tool. Behavioral VHDL is of course one of the user-input language of ASAR, and structural VHDL will be one of its possible output format. VHDL is used by the OSYS and GAUT synthesis tools.

The main concern of this paper is to describe the framework and the common formalism: the CENTAUR system is presented in section 3 and the data-flow graph format GC is described in section 4.

3 Centaur: a Distributed Programming Environment

3.1 Centaur

The CENTAUR system [4] is a generic interactive environment parameterized by the **syntax** and the **semantics** of programming languages. When provided with the description of a particular programming language—including its syntax and semantics—it produces a language-specific environment. With the help of CENTAUR one can develop tools needed for a programming environment: structure editors, type checkers, debuggers, interpreters, compilers, various translators and program transformations. These tools (generated from a formal definition of the language) have graphical user interfaces. For example, a type checker can generate a special editor with a list of error and warning messages; clicking upon any item will show the location in the source program that provoked the message.

The CENTAUR system has recently evolved in a set of cooperative tools, whereby we can couple CENTAUR with existing tools, e.g., an editor, a parser, or a compiler. This has two benefits: the generated system is no longer monolithic (different generated tools may run on different processors) and, more importantly, one can integrate existing tools (developed independently of CENTAUR) in the generated environment. Actually the system can be seen as a toolkit that helps in developing tools for manipulating programs or other structured objects, as well as a multi-formalism **framework** providing communication facilities and an homogenous user-interface to various independent tools.

The cooperation of tools implies communication and the sharing and exchange of data between these tools. In the following section, we present SOPHTALK [10, 11], an implementation of such a communication platform.

3.2 Sophtalk

Recent progress in software engineering techniques has consolidated the idea that any tool that performs a task (editor, button, typechecker, etc.) should be able to operate in isolation, without having to worry about sources of information it requires or recipients of information it generates [5, 6]. Rather than converse directly with its comrades, a tool announces an important event by broadcasting an appropriate message "into the world." The term *multicast* is more appropriate since a message only reaches the ears of those tools that have already declared their interest in such a message. Listening tools may react to a message, perform some activity, and emit their own messages. If no tools listen to a given message, it falls on deaf ears.

This event model of communication has been widely used to coordinate communication between several processes, but the principle of separating a tool's task from its communication needs also applies to single process tool interaction. A tool emits a message whenever it requires information, or when it accomplishes (or fails to accomplish) an important task. "I have finished compiling," "Someone has modified this file," or "The current selection has changed" are examples of significant events. Events may occur at any moment, so this model of communication is naturally asynchronous, but one may impose synchronization through design choices.

With SOPHTALK, one may design a system as a network of event-conscious tools. This kind of design allows developers to add, modify, substitute, or remove tools from the system without disturbing global behavior.

SOPHTALK implements an event model of communication. System objects, called *stnodes*, emit messages when significant events occur, such as the termination of a computation, a request for a service from another object, error conditions, etc. Messages circulate asynchronously in a network of stnodes. An stnode's type determines which messages it will receive; upon reception, an action corresponding to the message and stnode instance is triggered. An example of *stnodes* is illustrated by the figure 2.

A non-atomic node.

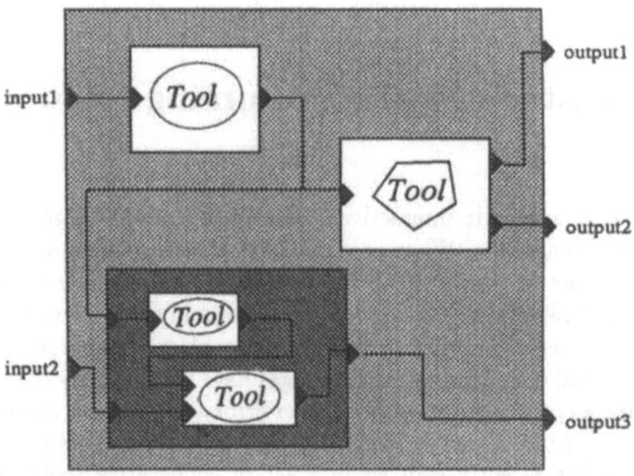

Figure 2: *Atomic and non-atomic stnodes*

3.3 Connecting Centaur with external tools

Some examples

Several projects have experimented with CENTAUR and connecting it to external tools using SOPHTALK. These include work around scientific computing, symbolic computation, and theorem provers, to name a few. One of the major achievements is a bi-directional communication between the environment and the symbolic tools allowing developments of interactive proof building, tactics designing, symbolic computing etc. In particular, CENTAUR can be used to provide a graphical user interface to a number of tools in the area of symbolic computation. CENTAUR communicates via a protocol with a number of computer algebra system (e.g., Maple, Sisyphe, Ulysses), with plotting/surface drawing tools (e.g., Gnuplot, ZicVis), and with theorem provers (e.g., HOL, Isabelle, Coq). Here one profits from the user interface of CENTAUR, and, for example, in the case of theorem provers one can take advantage of many techniques from programming environments: selecting the next subgoal to attack, saving an incomplete proof, browsing the various theories available, choosing tactics; the CENTAUR technology provides the Virtual Tree Processor, efficient pretty printing, and the generalized notion of location together with its tools for building a user interface.

3.4 Connecting tools within ASAR

In the ASAR project, some tools have already been developed under CENTAUR:

- OSYS: an environment for **VHDL**
- TRANSE: for LUSTRE program tranformations
- ALPHA DU CENTAUR: an environment for the ALPHA language.

A structure editor (parser and pretty-printer) has also been developed using CENTAUR for the SIGNAL language. The ASAR common framework built with CENTAUR allows these various tools to easily communicate, and to compare their results obviously. Moreover this common framework facilitates the design of the Internal Common Formalism **ICF**.

Furthermore CENTAUR's new architecture allowing integration of existing external tools is also useful in the ASAR context:

- communication between closely related tools as OSYS and GAUT is facilitated;

- communication with the external SIGNAL compiler makes the existing CENTAUR environment up (since a syntax editor already exists);
- communication with MATHEMATICA makes the cooperation between the two ALPHA DU CENTAUR's implementations (Centaur and Mathematica) easier;
- communication with a graph server allows computed graphs visualization.

Note that the two first communication (between Osys and Gaut, and between the CENTAUR framework and the SIGNAL compiler) will use the ICF format to transfer data. This Internal Common Format is also used to translate the framework results into the external tools input languages (structural or synthesisable VHDL, EDIF, C, etc.).

4 An Internal Common Formalism for ASAR: GC

As said before, it is expected that different tools and formalisms will be used or at least experimented for the same design, using ASAR. In order to facilitate this multi-formalism approach, the role of the ICF appears essential:

- as a target for synthesizing and transforming tools: input-L → ICF,
- as a source for translators into languages, like structural VHDL or C, which could be inputs for other external tools: ICF → output-L,
- as source and target for possible new transforming tools: ICF → ICF.

GC, as a synchronous data-flow format, is presumed to be the ICF for all the participants to the ASAR project. In this section, we present the place of GC in the common formats of synchronous languages, its main features and functionalities, and the tools already implemented applying to it.

4.1 The common formats of synchronous languages

GC was defined as one of the common intermediate formats of synchronous languages. The development of these formats has several major objectives:

- Establishing a standard for synchronous programming, the formats would support the largest part of the compilation process of all synchronous languages.
- Developing targeted user interfaces to the formats should be easy and of low cost.
- Developing targeted code generators for particular architectures should also be easy and of low cost.
- Providing tools and services within the synchronous technology (proofs, simulations, performance evaluation...) could be performed by different companies.

A first public version of the definition of the formats has been issued in june 1993 [15]. The formats are now a major component of the European SYNCHRON Eureka project, grouping together european companies and research centers involved in real-time studies and developments. A crucial issue of this project will be the standardization of the formats.

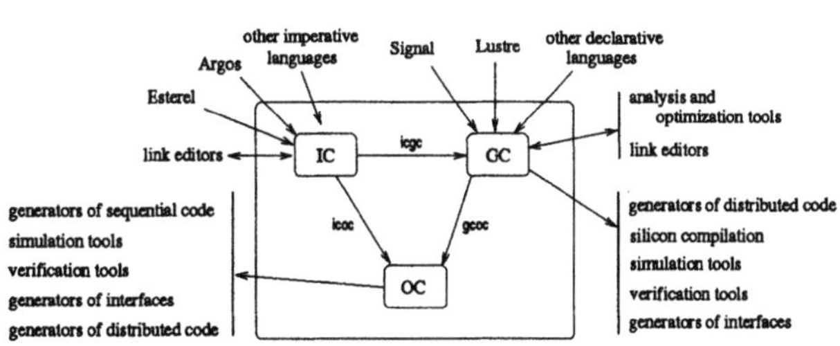

Figure 3: The Common Base for Synchronous Programming

The common formats consist of three formats relying on the same structures :
IC, a parallel format of imperative style.
GC, a parallel format of data-flow style, targeted in particular to LUSTRE and SIGNAL.
OC, a sequential format to describe automata.

A general view of the formats, expected translators between the formats, and planned relations with different tools is depicted in figure 3.

It should be noted that the word "format", used in contrast with "language", means that this object will not be accessible to end users but is rather designed to be handled easily by computers (neverthless, a readable decompilation such as that used in this paper can be very useful...).

4.2 Main common features

Main structures: As GC is a format (compared to a language), its syntax is strongly structured: it has been designed to be well accepted by automatic analysers. A **program** is a list of **packages**, which are lists of **entities**, themselves made of **tables of objects** (various entries of the tables). These structures are shown in figure 4. Packages do not have any semantics: they are used to put together entities linked by contextual meanings. They are referred to by their identifier; all other objects are referred to by indices, which avoids complex syntactical research of identifiers.

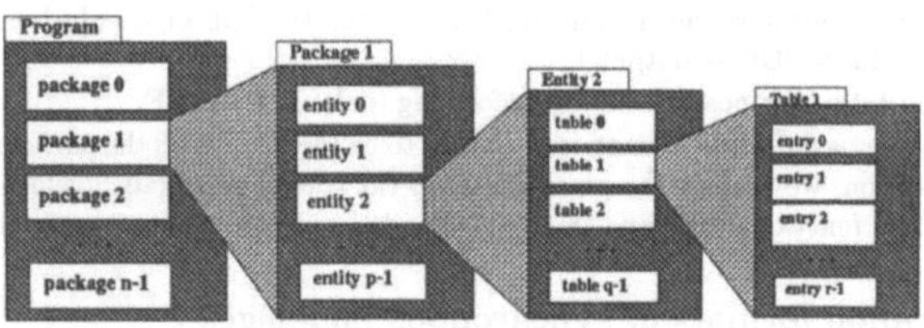

Figure 4: The main structures of the three formats

Comments: Though the formats are not designed to be read directly by users, the presence of comments, that may be useful for debugging, is allowed by the grammar. An other way to put comments in programs is the use of the pragma **%comment:%**.

Pragmas: Pragmas are "executable comments", kept by tools like translators between formats, and adding informations to objects to which they refer. At the moment, two pragmas are accepted by GC tools (especially the compiler): **%main:%** and **%comment:%**. The first one applies to a node (see 4.3), meaning that this one is to be considered as the principal node of a program. The second one can be found anywhere in a program. Pragmas appear as the easiest way to make GC evolve, once the main structures will be improved by the implementation of the first tools.

Addressing mechanism: As said before, the formats, designed for easy automatic analysis, own a very simple addressing mechanism. There are three ways to refer to an object:

- with a prefix, if the object belongs to common or GC **predefinitions** (see below). The prefixes are respectively **$** and **$g**.
- directly, by the index of the object, if it belongs to the current entity.
- indirectly, if the object belongs to an other entity. This other entity has to be imported by the current entity, by appearing as one entry of the table of **importations**. The index, called then *coumpound-index*, is made of the index of the corresponding importation, followed by the local index of the objet in the imported entity.

With this mechanism, all the objects can be reached by simple translations in packages and entities.

Entities: They are made of tables (distinguished one from the other by a keyword), and each of them can hold a table of importations. Importations are used in an entity to refer indirectly to objects declared in other entities. GC entities can be data blocks, interfaces, local or external nodes. Data blocks can hold tables of *types, constants, functions* and *procedures.* Those objects are only declared, which allows their use (their definitions are external and specified in an other way).

Predefinitions: Many applications use simple types, built-in in most languages. Those types are declared in a special data block called ***common predefined data block***. They are **`$pure`** (clocks), **`$boolean`** (logicals), **`$integer`**, **`$string`** (strings of characters), **`$float`** and **`$double`** (extended floats).

The predefined data block also declares constants: **`$top`** (always-present clock), **`$true`** and **`$false`** (always-present logicals). It also provides the most used functions on predefined types. This data block does not need to be imported by entities: common predefinitions are supposed to be known in all the programs written in any of the three formats. Any entity can refer to predefined objects without importing the predefined data block.

4.3 A brief overview of GC

GC is a hierarchical representation in the data-flow or block diagram style of synchronous programs, augmented with explicit control. The declarative part of a GC code has a synchronous semantics, presented in [15].

Since GC is a data-flow format, the basic object it deals with is the **flow**. A flow is a possibly infinite sequence of typed values with an explicit associated **clock** denoting the instants at which values are present.

Structure of a GC code A GC code is made up of a list of declarations of three kinds of entities: data blocks, **interfaces** and **nodes**. A node represents a declared sub-graph. It can be instanciated or activated (see below) with the only knowledge of its interface. There are two kinds of nodes: *local* ones and *external* ones. External nodes are nodes for which the GC description is not avaible. In particular, they can be the GC perspective of an imperative module. This is to be the link between GC and IC codes.

Interfaces The interface of a node is the summary required for use of the node as a "black box" in its instanciation context. It is made up of:

- **a data part** which describes mainly in form of data block imports, the types of flows entering and exiting the node (called *interface flows*), and, in the form of flow declarations, the interface flows and their clocks;
- **dependencies** which enable the specification of a partial communication order among the interface flows within the same instant;
- **assertions and synchronizations** which are the properties expressed on the interface flows.

Local nodes A GC local node has an interface, data and a body.

- As seen before, the interface of a node describes the properties which are visible from the outside.
- The data describe, in the form of **data block imports**, the types, constants, functions and procedures used by the node, and in the form of **flow declarations**, the local flows (the knowledge of which is not necessary outside of the node).
- The body of a node is made up of the following items:

  Definitions of values of flows: They express the computation of values of output (and local) flows in terms of functions of input (and local) flows. They are deterministic and obey the referential transparency principal (their lefthand side can substituted for the righthand side in all cases). The definitions can be *equations*, *node instanciations*, *procedure calls* or *activations of nodes*[2].

  A set of definitions can be seen as a data-flow graph. The vertices of the graph are the definitions. The edges are implicitly represented by the identity of indices of connected flows. The dependencies between the input flows and the output flows of a vertex are conditioned by the clock at which they are effective. For the vertices that are equations, these dependencies are induced from the operators used in the expression of definition; they are implicit, but they can be explicited in a table of dependencies (see below). For the vertices which are node instanciations or activations, the dependencies can be specified in the interface of the node. Moreover, explicit dependencies can be added between definitions.

[2]Several syntactic occurrences of instanciation of the same node produce distinct objects. At the opposite, several syntactic occurrences of activation of the same node make reference to the same object.

Synchronizations: They specify constraints governing the semantics of the program.

Assertions: They are properties expressed on the flows of the current node. They are not supposed to be computed by all tools.

Dependencies: They allow the specification of a partial execution order between communication flows in any instant.

An example of GC program is shown on the left side of figure 5; it is the translation of the data-flow graph obtained by the compilation of a SIGNAL program describing a FIR. An equivalent representation of the same program expressed in VHDL is shown on the right part of the figure. In this example, loops have been unfold, as it is done in GAUT for instance.

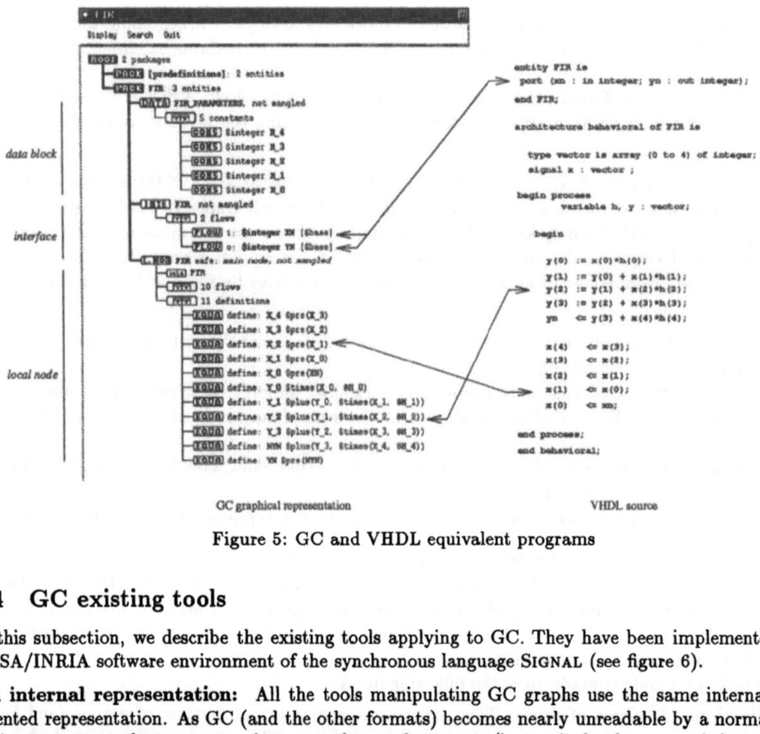

Figure 5: GC and VHDL equivalent programs

4.4 GC existing tools

In this subsection, we describe the existing tools applying to GC. They have been implemented in the IRISA/INRIA software environment of the synchronous language SIGNAL (see figure 6).

An internal representation: All the tools manipulating GC graphs use the same internal object-oriented representation. As GC (and the other formats) becomes nearly unreadable by a normal human brain as soon as the programs size more than a few pages (in particular because of the addressing mechanism), the associated decompiler produces two different files: the first one is written in GC as defined by the grammar, and the second one is a readable version where all the indices have been replaced by the corresponding identifiers.

A graphical browser: Useful for debugging and producing documents, the browser [3] becomes rapidly quite indispensible to GC users. In the future, new features will be added to the existing ones: multi-windowing, editing (copy/cut/paste) operations, PostScript output...

A SIGNAL to GC translator: An essential work of the SIGNAL compiler consists of a clock and dependency calculus; the result is a conditioned hierarchical graph (CHG). The CHG is a clock hierarchy organized as a forest of trees, where each clock owns a synchronous data-flow graph. The translator, integrated in the SIGNAL environment as an option of the SIGNAL compiler, builds a GC graph representing the clock hierarchy. The GC program presented in figure 5 has been produced by the SIGNAL compiler. The translator is unidirectionnal (from SIGNAL to GC), but the first experimentations reveal GC as a powerful representation and the format could become in a near future the intermediate code of the different phases of the SIGNAL compiler.

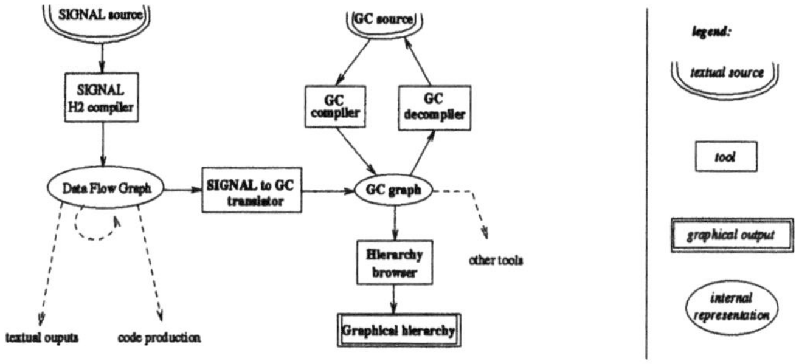

Figure 6: GC tools in the SIGNAL environment

A compiler: The global process of the compilation of GC is depicted in figure 7.

An intermediate phase of creation of an abstract tree from the source code has been introduced to enable the connection with other applications, without running the lexical and syntactical analysers.

The main problem of the compilation is the resolution of references, due to the addressing mechanism. This resolution is necessary to increase the security of tools applied to GC. This phase consists in attaching to each object the objects it may need in operations on the graph. The main steps are the following: *resolution of the importations,* where importations are attached to their corresponding entity, *resolution of the type-references,* where typed objects (constants, flows, functions, arguments of functions, parameters of procedures...) are attached to their type, and *last resolutions,* which consist essentially in the verification of all the objects referred to, and in type-matching.

The result of all the resolutions is an internal graph where the importations, only needed by the addressing mechanism, can be suppressed. The data-flow graph produced can be used for transformations, code production, optimisations, hardware synthesis, etc.

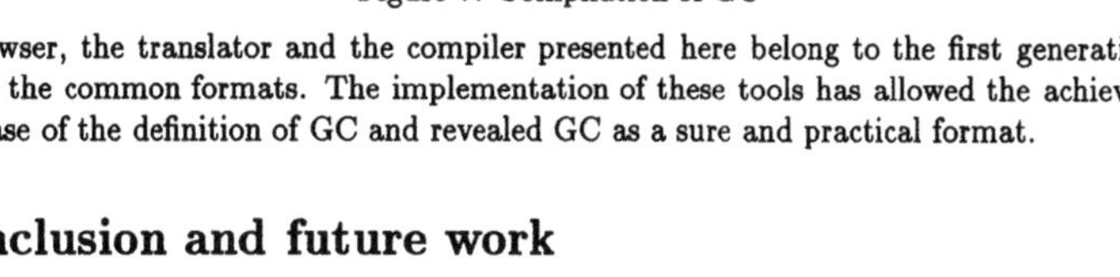

Figure 7: Compilation of GC

The browser, the translator and the compiler presented here belong to the first generation of tools applying to the common formats. The implementation of these tools has allowed the achievement of a second release of the definition of GC and revealed GC as a sure and practical format.

5 Conclusion and future work

Two main objectives have been defined in this project:

- integrate the different architectural synthesis tools in a generic environment,
- define an internal common formalism (ICF).

The first objective has been obtained with the implementation of Asar.0 which includes the system manager of the different tools in the CENTAUR environment. The designer disposes of a coherent framework, where different architectural synthesis tools co-exist.

For the second point, the data-flow graph format GC, with possible evolutions, seems well suited to play the important role of inter-communication between formalisms in our framework. Moreover, it will be used as a link to external tools such as generators of distributed code, formal verification tools, etc. It has now to be integrated as a possible input or output format in each one of the tools available in the ASAR project.

References

[1] *IEEE Standard VHDL Language Reference Manual.* IEEE, 1987. IEEE Std 1076-1987.

[2] P. ASAR. Towards a Multi-formalism Framework for Architectural Synthesis: the ASAR Project. In *Third International Workshop on Hardware/Software Codesign,* Grenoble, France, IEEE Computer Society Press, September 1994, 25–32.

[3] P. Aubry and S. Machard. *Représentation graphique d'arbres sous X11R5: Implémentation générique orientée objet, Applications.* IRISA, Rennes, to appear.

[4] P. Borras, D. Clément, Th. Despeyroux, J. Incerpi, G. Kahn, B. Lang, and V. Pascual. Centaur: the system. In *Proc. of ACM SIGSOFT '88: Third Symposium on Software Development Environments,* Boston, November 1988. (Also as INRIA research report number 777).

[5] D. Clément, A distributed architecture for programming environments. In *Proceedings of the Fourth ACM SIGSOFT Symposium on Software Development Environments,* Edited by R. N. Taylor, Irvine, California, December 1990, 11–21.

[6] D. Clément, F. Montagnac, and V. Prunet, Integrated Software Components: a Paradigm for Control Integration. In *Proceedings of the European Symposium on Software Development Environments and Case Technology,* Konigswinter, Germany, June 1991.

[7] G. Durrieu and M. Lemaître. Design by transformation of synchronous descriptions. In *3rd International Workshop on Algorithms and Parallel VLSI Architectures,* Leuven, Belgium, August 1994.

[8] N. Halbwachs, P. Caspi, P. Raymond, and D. Pilaud. The synchronous dataflow programming language LUSTRE. *Proceedings of the IEEE,* 79(9):1305–1320, September 1991.

[9] M. Israël and J. Benzakki. Osys: synthesis tool and hardware/software codesign. In *IFIP Codes/cashes93,* IFIP, Austria, May 1993.

[10] I. Jacobs and J. Bertot, *Sophtalk tutorials.* INRIA Rapport Technique no 149, Feb. 1993.

[11] I. Jacobs, F. Montagnac, J. Bertot, D. Clément, and V. Prunet, *The Sophtalk Reference Manual.* INRIA Rapport Technique no 150, Feb. 1993.

[12] P. Le Guernic, Th. Gautier, M. Le Borgne, and C. Le Maire. Programming real-time applications with SIGNAL. *Proceedings of the IEEE,* 79(9):1321–1336, September 1991.

[13] H. Le Verge, C. Mauras, and P. Quinton. The ALPHA language and its use for the design of systolic arrays. In *Journal of VLSI Signal Processing,* 3:173–182, 1991.

[14] E. Martin, O. Sentieys, H. Dubois, and J.L. Philippe. An architectural synthesis tool for dedicated signal processors. In *EURO-DAC 93,* 1993.

[15] J.-P. Paris, G. Berry, F. Mignard, Ph. Couronné, P. Caspi, N. Halbwachs, Y. Sorel, A. Benveniste, Th. Gautier, P. Le Guernic, F. Dupont, and C. Le Maire. *Projet SYNCHRONE – Les formats communs des langages synchrones.* Technical Report 157, INRIA, June 1993.

10

Enhancing the Data Openness of Frameworks by Database Federation Services

Ralf Böttger, Yaron Engel, Gerd Kachel, Silvia Kolmschlag, Dietmar Nolte, Elke Radeke
Cadlab - Joint Venture University of Paderborn and Siemens Nixdorf Informationssysteme
Bahnhofstr. 32, 33102 Paderborn, Germany
dbf-ma@cadlab.de

Abstract

Tools of existing frameworks can easily share data between each other by its integrated database system but data access to external database systems, file systems, or other frameworks is often not supported. We propose to incorporate database federation services into a framework to enhance the data openness of frameworks. These services allow a coupling of the frameworks database with external database systems while the autonomy of the individual database systems is preserved. Thereby, existing framework or database applications can continue running without any change. On the other hand, framework applications are enabled to access data of multiple databases uniformly and transparently via the old, possibly extended framework interface. This paper introduces the architecture of the proposed database federation services and presents two alternatives of their incorporation into a framework. The practical relevance is shown in an industrial environment where a framework is opened to the data of another framework and a PC database.

1 Introduction

Over the last several years, more and more CAx[1] systems use the benefits of a framework. All applications and tools of these systems can access the homogeneous user interfaces of the frameworks and exchange their data using the integrated database system (DBS) [4, 6, 9]. But the introduction of a framework is often handicapped by an empty or not sufficiently filled internal framework database system (FW DBS). Instead, the necessary data of the framework tools are already stored in file systems or other isolated, non-integrated database systems (see Fig. 1). The lack of data import facilities and database coupling services may impede the initialization of the internal framework database with the necessary data. Furthermore, different application areas in an enterprise often use different frameworks specific for their needs, e.g. CASE applications use a CASE framework. Each of these frameworks can have its own integrated DBS. Hence, there is a multitude of different, mostly isolated DBS in current enterprises. Although applications within a specific framework can share data, usually this is not possible in a controlled way to applications accessing other database or file systems.

[1] CAx is a place holder for CAD, CASE, CAM, CAE, EDA, etc.

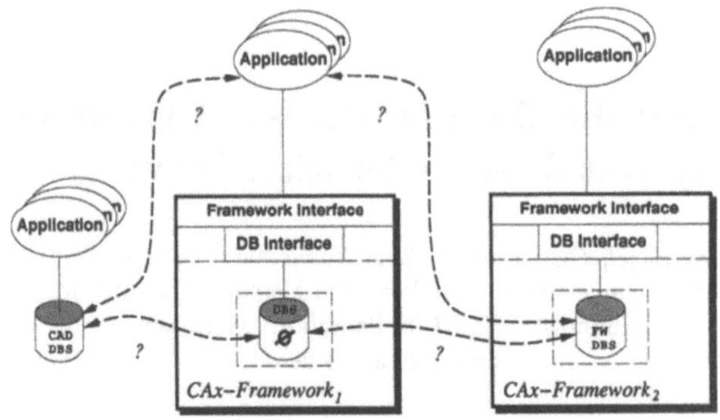

Figure 1: Frameworks lacking data openness

The introduction of database federation services (DBF) into a framework solves these problems and enhances the data openness of a framework. DBF offers a uniform, transparent access for framework applications to the FW DBS and to other DBS outside the framework. The chosen DBF architecture is very flexible and adaptable to the users needs. This is achieved by a modular, object-oriented design and generic software components.

In contrast to coupling techniques like import/export mechanisms or "gateways", DBF controls consistency across all coupled database systems. Furthermore, an extension of the coupled DBS is supported naturally by DBF but not by the other techniques. This good extensibility is achieved by the usage of a canonical data model within DBF and a generic adapter technique. Finally, we enhance the concepts of federated database systems [11] by different kinds of incorporation of DBF into a framework.

In Section 2, we present the DBF architecture and give an overview of its functionality. Its incorporation into a framework is explained in Section 3. The benefits of such an incorporation are stated in Section 4. To verify our concepts, we have started a pilot project with one of our industrial partners described in Section 5. Last, we summarize our paper and give an outlook to the future work in Section 6.

2 Database Federation Services

In order to exchange and share data with other frameworks, database systems, or file systems, they have to be coupled somehow. There are two alternative architectures to achieve this (Fig. 2): (1) a coupling layer for each data system realizes the mapping to the others or (2) a single coupling layer with a canonical data model maps to all data(base) systems back and forth. In both alternatives, users can access the data system directly via its interface or use the coupling layer in order to access data of multiple/other data systems.

The first alternative requires $n * (n-1)$ mappings between data systems with n the number of data systems while the second alternative needs n mappings. But current enterprises have a multitude of data management systems, so n is big (e.g. 20 for a department of an industrial project partner). Moreover their distribution, heterogeneity, and autonomy pose many tasks to the mapping, e.g. data model and schema transformation as well as redundancy control [7]. Therefore, we chose the second alternative for our architecture in order to decrease implemen-

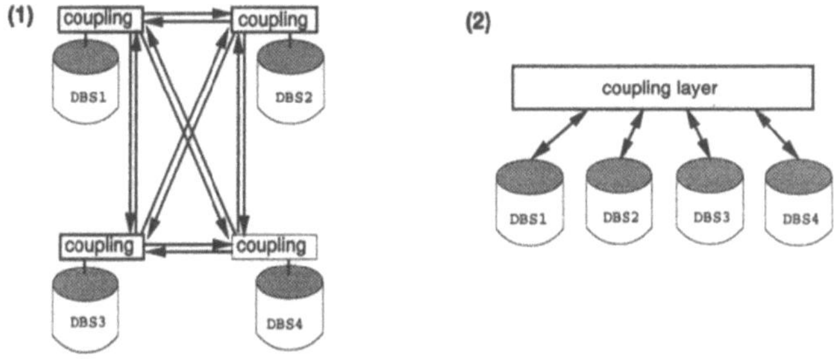

Figure 2: Alternative architectures:(1) point-to-point-connections, (2) canonical coupling layer

tation effort and ease extensibility with further database systems.

Fig. 3 presents our architecture in more detail. It represents a federated database system [11] preserving the autonomy of the distributed heterogeneous DBS. The DBS specifics are encapsulated in the DBS adapters while all other modules and interfaces of the database federation services are generic. When installing DBF in a specific environment, only DBS adapters for the enterprise's data systems have to be provided. The *DBS adapters* realize the communication with the various coupled database and file systems of the enterprise, e.g. with hierarchical, network, relational, and object-oriented DBS, or file systems. Moreover, they transform the heterogeneous data and operations into a canonical data model. For this, a coupling interface with 14 operations such as *createObj, deleteObj, loadObj, saveObj* is mapped to corresponding DBS data and operations. To give an impression of their complexity, a DBS adapter for relational DBS with an SQL interface needs about 3000 lines of C++ code (including comments) whereas that for an object-oriented DBS [6] requires only about 600 lines of C++ code. To ease the dynamic extension of the federation with new data(base) systems, there is a uniform layer (*database coupling*) on top of the DBS adapters.

As canonical data model, we use an object-oriented data model due to its expressive power [10]. It is based on the upcoming standard for object oriented database systems, the ODMG object model [2] which provides concepts like objects, classes, class inheritance, and object identity. We extend the ODMG data definition interface (ODL) by functionality for schema mapping/integration. Data are accessed either locally by the DBS interfaces of the autonomous data systems or globally via *global interface* which is based on the ODMG C++ OML. It allows a uniform and transparent access to data of all coupled DBS. Other interfaces are realized as *external adapters* on top of it, e.g. application specific interfaces or standard interfaces like SQL. In contrast to the DBS adapters, the realization of an external adapter is more expensive because the complete database interface has to be implemented and not only a subset. Nevertheless DBF will offer external adapters for standard database interfaces and only specific external adapters would produce extra development effort.

By the *administration interface*, the DBF is started, stopped, initialized, and DBS are coupled/decoupled to/from the DBF. The administration services are realized by three modules: *process management, coupling control,* and *user administration* (see Fig. 3). The user administration is responsible for the mapping from global DBF user to local DBS users. Beside these administration modules, the *federation kernel* contains several other modules which realize tasks such as query decomposition, global transaction management, and global redundancy control. A central role among these modules plays the *object manager* which is based on the context object

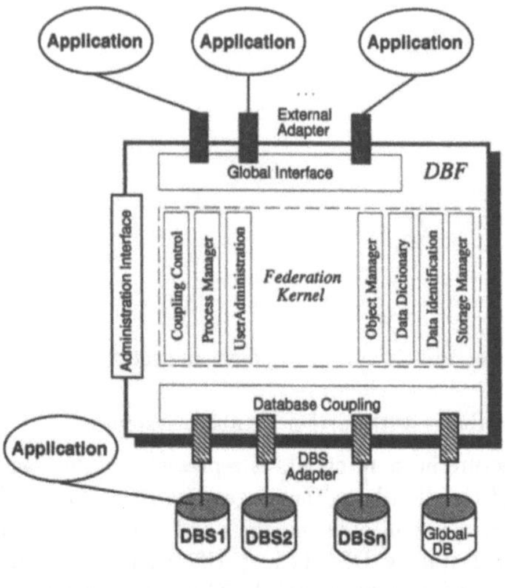

Figure 3: DBF architecture

model [5]. The object manager translates the global object accesses into local DBS objects (but still in the canonical DBF data model) and vice versa. These translations are done according to the 5 level schema architecture for schema integration in federated database systems [11]. The *object manager* uses the *data dictionary* module which stores the schema integration information and the *data identification* module which realizes the mapping of global resp. internal object identifiers to the heterogenous local DBS identifiers. The module *storage manager* is responsible for an efficient transfer of data between main memory and persistent storage including the caching of local DBS data.

Schema and configuration information as well as global new data like global relationships that do not map to any DBS are stored in an auxiliary database system. This *global DBS* is coupled via a DBS adapter.

3 Incorporation into a Framework

In this section, we present some new ideas for the incorporation of database federation technology into a framework. Figure 4 shows two possible alternatives how DBF can be incorporated into a framework. In both alternatives, DBF couples the internal framework DBS with other (external) DBS via the DBS adapters in a controlled way. Furthermore, a specific external adapter offers the complete framework database interface and manages the mapping between the framework data model and the canonical DBF data model. In the remaining section, we discuss both incorporation alternatives.

DBF as External Service

Alternative (1) couples the FW DBS in the classical way [11] whereby all coupled DBS are remaining autonomous. I.e. the existing applications and systems are not affected by the DBF (no relinking and no performance loss). But this architecture requires an already open FW

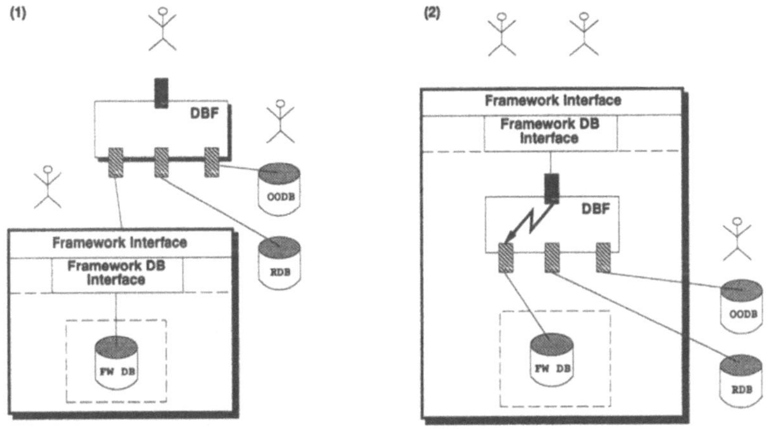

Figure 4: Two possible incorporations of DBF into a framework: (1) external and (2) internal

DB interface. I.e. the framework has to offer the database interface as a separate service which is not encapsulated into other framework services. This alternative allows a very easy introduction of DBF into an existing environment using framework technology. Neither the existing framework(s) nor their (local) applications have to be changed. On the other hand, new (global) applications can access the federated data of different systems coupled by DBF. But there are consistency problems between the local and the global applications [1], e.g. only *quasi-serializability* can be guaranteed [3].

DBF as Internal Framework Service

In contrast to alternative (1), alternative (2) allows full consistency control for all framework applications. Therefore, any data access using the framework interface is passed through the external DBF adapter which includes a consistency check for all federated data.[2] But the general consistency problem between the framework applications and applications using only the external DBS still remains.

This alternative implies a change of the internal structure of the framework but the framework interface is stable. Thereby, existing applications could access (more) federated data via the same framework interface in a consistent way. This is also true if the set of coupled DBS changes e.g. by the coupling of a new DBS. A relinking of the existing framework applications becomes necessary if the incorporation of DBF into a framework affects the framework library which has to be linked by the applications. In a client-server architecture, where only the framework server changes through the incorporation of DBF, no relinking of the applications (clients) is necessary.

The passing of all data accesses via the external DBF adapter and the internal data model transformation may produce a performance loss for the complete framework. This is not acceptable for framework applications which are using data stored only in the FW DBS. To avoid this performance loss, unnecessary data model transformations between the framework database interface using the framework data model, DBF using the canonical DBF data model, and the FW DBS using again the framework data model are shortened. Therefore, the external adapter

[2] A detailed description of the consistency check will be presented in a future paper.

directly maps a data access to the internal FW DBS without using the DBF kernel if no other (external) DBS is affected by this data access. I.e. the DBF kernel becomes only active if a federated, external DBS is involved for the data access. These shortcuts of the external adapter allow an efficient processing of the original framework data. The DBF concept distinguishes between different kinds of shortcuts which are not discussed in this paper but we plan to prepare another paper with this topic.

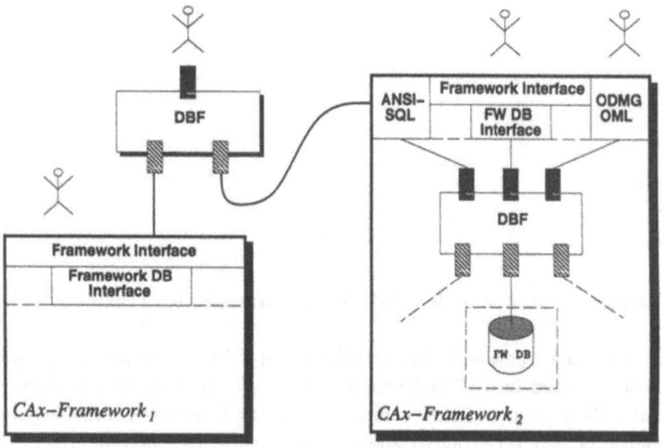

Figure 5: Coupling of different frameworks by using DBF

Beside enhancing the framework data access facilities by the federation of other (external) DBS, DBF can also enhance the openness of the framework to other frameworks or systems. This is realized by a further external DBF adapter, e.g. a standard database interface like ANSI-SQL or ODMG C++ OML. The "standard" external adapter allows a standardized data access to all federated DBS including the FW DBS. In alternative (2), this further external adapter implies an extension of the FW interface by a new "standardized" database interface (see Figure 5). The extended framework interface can be used by other applications which are not integrated into the framework or can be used for coupling the framework with another framework via a second external DBF.

This enhanced data openness of a framework is the first step in the direction of *framework interoperability*. DBF offers here a generic, powerful, and consistent solution.

4 Benefits

Openness is an important intention for any framework to support the exchangeability and integration of any tool or data. The integration, especially a deep integration, provides uniform access to shared data. For example, the integration of an expert system, which contains its own knowledge base system within a framework may provide access to shared data. The complete benefits of openness [9] can only be achieved through a standardized tool interface. The database federation services enable, via the global and external interface, the "standardized" access and interchange of heterogeneous data between tools within the framework and, nevertheless important, between tools of various frameworks as well.

The major benefits by using the database federation services are listed in the following:

- Existing framework applications or applications of the coupled data systems can continue running so that no data or application has to be migrated or recoded.
- Because existing database or file systems do not need to be replaced but can be supplemented by new ones, investments will be saved.
- The relaxation of the strong connection of applications to one or only a few special database systems allows an easy exchange and switch between different technologies.
- Common data do not have to be stored redundantly in different database systems or file systems.
- New framework applications can uniformly and transparently access data of the framework's integrated database system as well as other coupled database and file systems. Hence, applications crossing multiple frameworks can also easily developed.
- The database federation services control consistency across all coupled database and file systems. Thereby, consistency checks are not restricted to the framework's integrated database system. Moreover, consistency is performed also for operation sequences that access multiple databases in terms of global transactions.
- The framework user may define and access relationships among data of different database or file systems.
- The whole company can benefit because the various departments can easily share and exchange data which eases concurrent engineering as well as enterprise integration. Moreover, the possibility to realize one/few information models for the whole company resp. cooperation makes global plans and decisions much easier for the management.

5 Application in a Pilot Project

An important part during the realization of the database federation services is the verification of our concepts by a pilot project within a company where problems solvable by DBF can be found. Therefore, an application on base of a DBF pre-release will be realized within a department of Siemens Nixdorf Informationssysteme (SNI).

Within the SNI department, there are several heterogeneous systems for managing hardware component data. Therefore a CAD framework consists of several systems for handling CAx hardware component data and stores technical information like the component's technology the relational database ADABAS[3] offering an SQL interface. An electronic design framework realizes a development environment for compatibility checks of electro magnetical circuits and is supported by various tools and a data handling system based on file system.

Within this environment, many applications have to access data of different database systems or file systems. This had been realized by import/export functions supported by a cross reference list for identification of data. To ease this large-scale process, the department requested a system which offers coupling of both frameworks and allows access to all database systems and

[3]ADABAS is a trademark of SOFTWARE AG.

file systems. Therefore, the actual problem is the lack of efficient support for data exchange between these data handling systems of the frameworks.

The DBF pilot concerns the controlled data coupling of the CAD framework on a DEC workstation, the electronic design framework on a SUN workstation, and a relational database for personal computers, namely ACCESS[4], as a hardware component library where data of the two frameworks shall be stored for the usage of PC tools. For realization, we chose the first alternative discussed in Section 3 where DBF is an external service of the furtheron autonomous frameworks. The following data handling systems will be federated by DBF:

- relational database ADABAS storing CAx data of the CAD framework
- Unix file system library storing the hardware component data of the electronic design framework
- relational database ACCESS for PC storing hardware component data for the administration of component data of the electronic design framework.

The necessary data exchange and data conversions will be managed by DBF on a SUN workstation. The schemas of the file system and the relational database ADABAS will be unified and the resulting federated schema will be stored in the PC database. The administration of the electronic design framework uses the PC database to initialize the data exchange. Global relationships between electronic design components and CAD components will be stored in the internal DBF database. The federated data will be accessed via an external adapter of the *global DBF interface* the electronic design administration via remote shell.

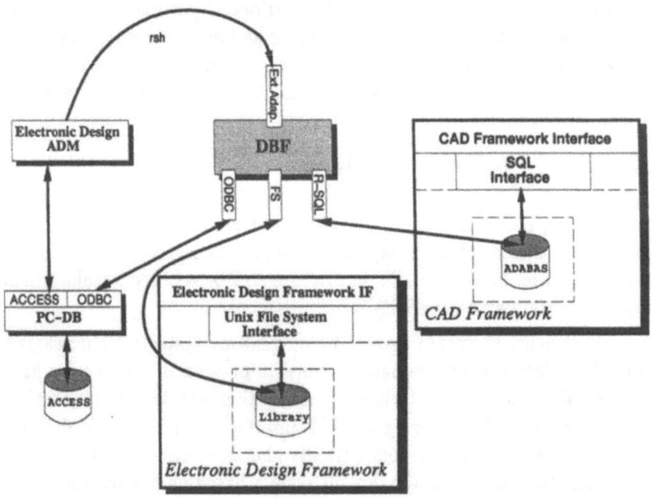

Figure 6: Data handling systems coupled by DBF pilot

For connections from DBF to the different data handling systems, different adapters have to be provided offering different interfaces to the data handling systems (Fig. 6):

[4]ACCESS is a trademark of Microsoft.

- A *Remote SQL adapter* realizes data exchange between ADABAS on a DEC workstation and DBF on a SUN workstation across hardware borders.
- A *file system adapter* offering a hierarchical oriented file format connects the Unix file system library storing electronic design data to DBF. The electronic design framework offers such a hierarchical oriented file format interface providing direct access to the file system library.
- An *ODBC adapter* ("Open Database Connectivity")[5] couples the PC database ACCESS to DBF. The usage of ODBC allows a direct coupling of the database without creation of temporary import/export files.

The concepts of federation and migration have been realized for the DBF pilot. Within the SNI department, the federation of the two frameworks provides a uniform and transparent interface to all federated systems. The migration of hardware component data supports the reduction of database and file systems and the realization of a single hardware component library. For administration of hardware component data, efficient data access will be realized and data consistency will be controlled automatically.

6 Conclusion

We showed the useful enhancement of frameworks by database federation services (DBF). These new services enable framework applications to access beside the internal framework database system also existing external database systems resp. other frameworks in a transparent and uniform way. This enhanced data openness of a framework is the first step in the direction of *framework interoperability*.

The presented DBF architecture is very flexible and adaptable to the users needs. This is achieved by a generic adapter technique between applications and DBF as well as between DBF and (autonomously) coupled databases. The DBF kernel is realized as a toolbox of configurable, object-oriented services. This can be used in different kinds of incorporation between DBF and frameworks. Our approach enables multiple kinds of framework incorporation and additionally, uses object-oriented implementation technology which allows itself federation of any data from flat relations to highly structured objects. This combination leads to a variety of benefits for framework users. Some of these benefits are reflected by our pilot application where concepts will be proven and our implementation will be improved.

Beside the pilot application, we are providing our concepts for the JESSI[6] framework JCF [6]. Since our first DBF release will provide basic functionality, it is obvious to enhance functionality in detail in future, e.g. more sophisticated transaction mechanisms and global recovery mechanisms. One of the more general extensions is the migration of data from one database to another as well as migration of applications (tools) from one database to another [8].

[5] ODBC is a trademark of Microsoft.

[6] This work is granted by the ESPRIT project JESSI COMMON FRAME No. 7364.

References

[1] BREITBART, Y., GRACIA-MOLINA, H., SILBERSCHATZ, A. Overview of multi-database transaction management, 1992.

[2] CATTELL, R.G.G. *The object database standard: ODMG'93.* Morgan Kaufmann Publisher, 1994.

[3] DU, W., ELMAGARMID, A.K. Quasi-serializability: a correctness criterion on global concurrency control in InterBase. In *5th Int'l Conf. on Very Large Data Bases (Amsterdam, Netherland)*, pages 347–355, 1989.

[4] European Computer Manufactures Association (ECMA). Reference model for frameworks of software engineering environments, 3rd edition, 1993.

[5] KACHEL, G. A multilayered database system architecture for supporting tool integration. In *Proc. of the 2. International IFIP WG 10.2 (Charlottesville)*, 1990.

[6] KRANNICH, G., DIPPER, H., FOX, W., KATHOEFER, T., SCHLEGEL, W., SCHMID, J. Jessi-Common-Framework V4.0, Global View, Detailed Functional Specification SP2. JCF Report JCF-SNI/055-01/29-Apr-94, 1994.

[7] NOLTE, D., KOLMSCHLAG, S. , KACHEL, G., RADEKE, E. Coupling Services for Heterogeneous Database Systems, Jessi-Common-Framework V4.0, Detailed Functional Specification. JCF/CADLAB/068-01/11-April-94.

[8] RADEKE, E., SCHOLL, M.H. Federation and stepwise reduction of database systems. In *Proc. 1st Int'l Conf. on Applications of Databases (Vadstena, Sweden)*, 1994.

[9] RAMMIG, F.J., STEINMÜLLER, B. Frameworks and design environments (in german). *Informatik Spektrum 15*, pages 33–43, 1992.

[10] SALTOR F., CASTELLANOS M., GARCIA-SOLACO M. Suitability of data models as canonical models for federated databases. *SIGMOD RECORD 20(4)*, pages 44–48, 1991.

[11] SHETH, A., LARSON, J. Federated database systems for managing distributed, heterogeneous, and autonomous databases. *ACM Computing Surveys 22(3)*, pages 183–236, 1990.

11

Interoperability Support by Integration of EXPRESS Models*

Gerhard Scholz and Wolfgang Wilkes
University of Hagen, Praktische Informatik I
Feithstr. 140, D-58084 Hagen
Germany
e-mail: {gerhard.scholz | wolfgang.wilkes}@fernuni-hagen.de

1. Introduction

Interoperability of various tools is a major goal of the development of frameworks and design environments. Tools are supposed to work cooperatively together, i.e. they have to work on common data and to exchange their results. Unfortunately, quite often different tools work on different internal models of the world, the framework's database may provide an additional model, and if more than one framework is involved, it is very likely that it does provide yet another model.

One attempt to solve this problem is the provision of standardized information models (usually formulated by means of EXPRESS [EXP92]) which are used as a basis for data transfer (e.g. STEP file format) or for database accesses (e.g. via SDAI). These efforts are very important, but it is doubtful that the ultimate description of a domain within one information model can be achieved. Different standards will remain existing in the future, and every CAD domain has to be open for new applications and new technologies which will require new data. Thus, we will have to deal with a variety of tool internal models, framework schemas and even standards. Supporting the interoperability of information models or integrating them sets an important basis for the interoperability or integration of the tools, frameworks, and standards themselves.

One fundamental prerequisite to support interoperability on the level of the information model is to clearly identify the commonalities and differences of the models under consideration. On the first glance this seems to be an easy task. But in reality, this requires a lot of effort since the models usually describe common concepts from different perspectives using different modelling styles. Furthermore, besides the difficulty to identify the common structures it is also necessary to describe these common structures in an unambiguous way. Considering these aspects, the idea was born to define a core model which exactly describes the commonalities of the underlying information models.

* This work was supported by the EC in the context of the ESPRIT project 8370 ESIP

A core model can only be of practical use if its relationships to the original models are described in a formal way. Only then it is possible to provide a common view of the overlapping areas of the models. Additionally, this sets the basis for the derivation of transformation programs which allow to transform data from one representation via the core model into another one. A further possible application could be the provision of an integrated view of the information content of both original sources (e.g. the integrated handling of VHDL and EDIF data).

The goal of this paper is to provide a methodology which allows to describe the common structures of different models and the way how they can be mapped to each other. Because we felt that the creation of a core model (schema intersection) and the creation of an integrated model (schema union) are strongly related problems, we started with inspecting the work which has been done in the area of schema integration for heterogeneous database systems (overviews can be found e.g. in [BLN86], [ACM91], and [IEEE91]). Our investigations showed that the models for electronic CAD comprise specific problems. Some of these problems are caused by EXPRESS since EXPRESS is not defined in an orthogonal way, i.e. a real world concept can be modelled in various ways. On the other hand, the models we are dealing with are far more complex than those considered in the heterogeneous database world. In particular, the entity types are much more interrelated, and it is not sufficient to only deal with single entities but we need to compare complex structures of entities and their relationships as a whole.

Our particular method has its roots in the concepts described in [SST92]. These give us a framework for describing the relationships of the core model to the original models in a formal way. In this paper, we will describe an adaptation of that method which has been tailored to the specific needs which arise in the engineering domains, more specifically which we were faced with during the process of developing a "core model" for EDIF and CFI. Some simplified examples from this core model [SchW94b] will illustrate our concepts. Here, we use *EDIF* always in the meaning of *EDIF version 3 0 0* [EDIF93]. Similarly, *CFI* is used as an abbreviation for *CFI DR version 1.0.0* [CFI92]. When refering to entities of the CFI model, we will omit the prefix "cfidr".

The rest of this paper is organized as follows: In chapter 2, the basic idea of the core model creation is described. In chapter 3, this approach is extended to the "7-schemas-model" which provides the description of mappings between the original representations via the core model. In chapter 4, the situation of just partial overlapping concepts is discussed. Chapter 5 deals with the consideration of constraints. Chapter 6 concludes the paper.

2. Creation of a Core Model

The basic idea of our method corresponds to the concept of views for multi-objectbases as described in [SST92]. There, a *same* function is used to describe that objects from different objectbases represent the same real world entity. To these objects - more correctly: to their classes - views are created to unify attributes and relationships and to solve semantic heterogeneity. Finally, a union view of these views will be created which covers the common concepts. The presentation in [SST92] uses COCOON as the underlying object-oriented data model, where view declarations can be directly integrated into the class lattice.

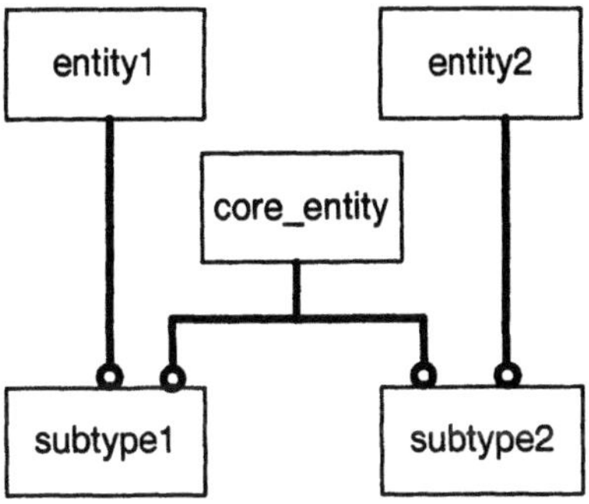

Fig. 1: Unification by a Core_Entity

Fig. 1 shows the modification of this concept for EXPRESS schemas. At first, a pair of entities - entity1 from schema A and entity2 from schema B - will be identified which represent the same real world entity. By the definition of views, different representations of attributes and relationships will be adapted to each other. In EXPRESS this is done by means of derived attributes, which will be declared in subtypes. This step does not add new information, only already inherent information will become visible. Therefore these subtypes are semantically equivalent to their respective supertypes. Finally, the common attributes of these subtypes will be included in a common supertype core_entity. Thereby the core_entity unifies entity1 and entity2.

Fig. 2 shows (simplified) excerpts from the information models of EDIF and CFI using the EXPRESS-G notation. In both models, a *cell* denotes an electronic chip, which can be described in different ways. A specific description is usually called *view*. In the following, we will consider just one view type, the netlist representation, which describes the electrical connections of the chips, called *NetlistView* in the CFI model and *connectivity_view* in the EDIF model. While CFI models this view by just one entity type, EDIF distinguishs between interface and implementation of the view. Furthermore, there exists the concept of a cluster, which gathers different views with the same interface.

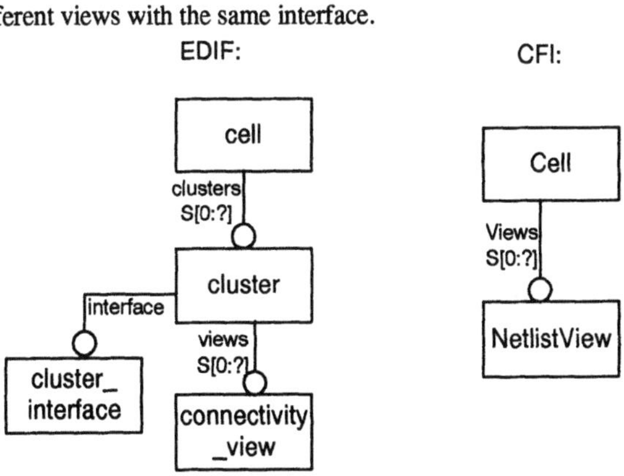

Fig. 2: *Cell* and *View* in EDIF und in CFI

Obviously both models include the concept of a *cell*, which - directly or indirectly - contains *views*. Exactly this information will be captured in the core model. This is shown in example 1, where *cm_connectivity_view* denotes the entity of the core model which unifies the concepts of *NetlistView* and *connectivity_view* (and *ee_..* and *ec_..* denote the corresponding subtypes[1]). Because the concept of the *cluster* has no correspondence in the CFI model, it will not be part of the core model. In the subtypes *ee_cell* and *ec_Cell*, it is described, how the core model information can be generated from an EDIF or CFI information base, respectively. This example already shows, that the main emphasis of our research is not the comparison of simple attribute types, but the analysis of complex entity structures and their relationships.

The choice of names for entities and attributes of the core model is arbitrary. Furthermore, even the structure of attributes can be chosen arbitrarily to some extend. Each equivalent representation of the core model structure is in principle possible. But usually the structure of the original models will be adopted as far as possible.

```
ENTITY cell;          -- Schema A (EDIF)
 clusters : SET [0:?] OF cluster;
END_ENTITY;
```

```
ENTITY Cell;          -- Schema B (CFI)
 Views : SET [0:?] OF NetlistView;
END_ENTITY;
```

```
ENTITY cm_cell;     -- Schema Core_Model
      views : SET [0:?] OF cm_connectivity_view;
END_ENTITY;
```

```
ENTITY ee_cell
SUBTYPE OF (cell, cm_cell);
DERIVE
  SELF\cm_cell.views :
   SET OF ee_connectivity_view :=
    all_views( clusters);
END_ENTITY;

FUNCTION  all_views(  sclu:  SET  OF
cluster): SET OF connectivity_view;
LOCAL
scv : SET OF connectivity_view:= [];
END_LOCAL;
REPEAT j:= 1 TO SIZEOF( sclu);
  scv := scv + sclu[j].views;
END_REPEAT;
RETURN( scv);
END_FUNCTION;
```

```
ENTITY ec_Cell
SUBTYPE OF ( Cell, cm_cell);
DERIVE
  SELF\cm_cell.views :
   SET OF ec_NetlistView
    := SELF\Cell.Views;
END_ENTITY;
```

Example 1

3. The 7-Schemas-Model

The actual goal in the definition of the subtypes is the description of the transformation from one representation (e.g. schema A) to another one (schema B). The derived attributes, however, describe just the mapping from the original schemas to the core model, but not the

[1] The prefixes cm_, ee_, and ec_ stand for "core model", "export EDIF", and "export CFI", respectively

other direction. The question remains open how to represent the core model data within the original schemas. In example 1, for instance, there does not exist a unique straightforward solution for this. *Cell* and *connectivity_view* are not directly related in EDIF, but only indirectly via *cluster*, which does not exist in the core model. Thus, additional decisions have to be taken to map from the core model into an original schema.

The simpler example 2 demonstrates that this problem occurs, if the mapping from the original schema to the core model is not reversible. Wheras the *PortInstScalar* in the CFI model has one *Props* attribute, the same information in the EDIF entity type *instance_port* is partitioned into two attributes *new_properties* and *overriding_properties*. The union of these set valued attributes leads from EDIF to the core model. To answer the question how CFI data can be represented in EDIF, also the direction from the core model back to EDIF needs to be defined. The additional specification of the mappings from the core model to the original schemas completes the 7-schemas-model.

```
ENTITY instance_port;    -- EDIF
      new_properties           :
            SET [0:?] OF property;
      overriding_properties    :
            SET [0:?] OF property;
END_ENTITY;
```

```
ENTITY PortInstScalar;   -- CFI
       Props : SET [0:?] OF Prop;
END_ENTITY;
```

```
ENTITY cm_instance_port;      -- Core_Model
      properties   : SET OF cm_property;
END_ENTITY;
```

```
ENTITY ee_instance_port
SUBTYPE OF  ( instance_port,
              cm_instance_port);
DERIVE
  SELF\cm_instance_port.properties
     : SET OF ee_property :=
            new_properties +
            overriding_properties;
END_ENTITY;
```

```
ENTITY ec_PortInstScalar
SUBTYPE OF  ( PortInst,
              cm_instance_port);
DERIVE
  SELF\cm_instance_port.properties
       : SET OF ec_Prop := Props;
END_ENTITY;
```

Example 2

In Fig. 3 the original schemas are named A and B and the core model is named C. The other schemas contain the subtypes which describe the respective transformation. The lines ending in a circle show, which schemas use information of other schemas. (EXPRESS offers interface specifications for this purpose.) The schemas A, B, and C can be used by themselves, whereas the transformation schemas, of course, need the information about source and target of the mapping. It should be observed, that at first with this approach the transformations to and from the core model can be defined independently of each other.

Example 2a complements example 2. For the mapping from the core model to EDIF, the *properties* of the core model will be attached to the attribute *new_properties*. Of course, this is just one of several possibilities. But it is essential that the chosen solution is documented within a formal frame.

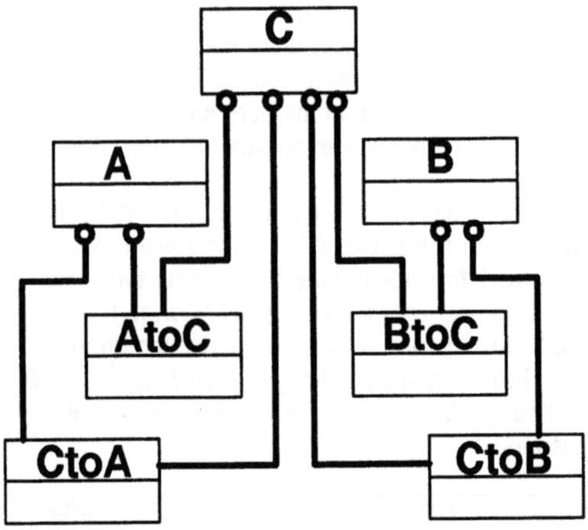

Fig. 3: The 7-schemas-model
(EXPRESS-G schema level diagram)

```
ENTITY ie_instance_port             -- Schema CtoA (with A = EDIF)
SUBTYPE OF  ( instance_port, cm_instance_port);
DERIVE
  SELF\instance_port.new_properties : SET OF ee_property :=
            SELF\cm_instance_port.properties;
  SELF\instance_port.overriding_properties      : SET OF ee_property := [];
END_ENTITY;
```

Example 2a

When data are transferred from an original schema to the core model and then back to the representation of the original schemas, they usually will not have their original shape. A transformation cycle via the core model changes *overriding_properties* to *new_properties*. That raises the question on the relationship between the mappings to and from the core model. Which information will be maintained, and which will be lost? The answer is obvious: the core model defines just the information which can be transferred without loss.

- **Rule 1:** A mapping from the core model into one of the original schema representations and back to the core model has always to be the identity function.

This restriction is not guaranteed by the 7-schema-model itself. But it is a minimal requirement for the correctness of the mappings and the legitimate use of the term core model.

A further requirement can be explained using example 1: when mapping from the core model to EDIF, the core concept *cm_connectivity_view* will be mapped not only to the EDIF *connectivity_view*, but some attributes will be mapped to *cluster* and to *cluster_interface* (fig. 2). Therefore schema CtoA (with A = EDIF) contains 3 subtypes of *cm_connectivity_view*, which describe the mapping of certain attributes to the corresponding EDIF concept. EXPRESS facilitates this way of description, because an object of a supertype may belong simultaneously to several subtypes.

- **Rule 2:** When mapping from the core model, structural refinement may take place. The other way around, when mapping to the core model, such a partitioning of a concept of an original model must not occur. This would contradict the conception of a core model as an intersection model.

4. Partial Mapping of Concepts

Up to now, in the examples it was assumed that either no or all objects of a certain type can be mapped. But an essential feature of EXPRESS is the declaration of constraints which may be different in the different models. Not only explicit integrity constraints, but also the structure of the models itself restrict the set of objects, which may have a correspondence in the other model.

An extension of the approach described above allows mapping as well, when the concepts of the original schemas do not correspond completely, but just overlap. In those cases, constraints will be attached to the respective subtypes which describe the mapping to the core model. These constraints will restrict the mappable set of objects in an appropriate way.

```
ENTITY instance;  -- EDIF
  instantiated_cluster : cluster;
END_ENTITY;
```

```
ENTITY Inst;        -- CFI
  Describer : NetlistView;
END_ENTITY;
```

```
ENTITY cm_instance;      -- Core_Model
      describer    : cm_connectivity_view;
END_ENTITY;
```

```
ENTITY ee_instance
SUBTYPE OF (instance, cm_instance);
DERIVE
  SELF\cm_instance.describer :
      ee_connectivity_view :=
SELF\instance.instantiated_cluster.views[1];
WHERE
      SIZEOF(
  SELF\instance.instantiated_cluster.views)
       = 1;
END_ENTITY;
```

```
ENTITY ec_Inst
SUBTYPE OF
       ( Inst, cm_instance);
DERIVE
 SELF\cm_instance.describer
   : ec_NetlistView
   := SELF\Inst.Describer;
END_ENTITY;
```

Example 3

The contents of example 3 relates to example 1. The implementation of a view mainly uses and connects views of other cells. The use of a view within the context of another view is in the area of electronic design usually represented by an object of its own, called *instance*. In CFI, this fact is described by an entity type *Inst* with an attribute *Describer*, which points to the used *NetlistView*. EDIF has a more general concept and allows additionally the use of incompletely described views. An interface description is sufficient, which is supplied by a reference to a *cluster*. Because CFI - and therefore also the core model - does not know the concept of a cluster, arbitrary instances cannot be mapped to the core model. But just to claim that the concepts are not mappable at all, would not do justice to the situation. A mapping is possible when the *instantiated_cluster* contains exactly one *connectivity_view* (fig. 2). In example 3 exactly this constraint is declared in the entity subtype *ee_instance*. In that case together with the cluster a unique view is specified, and the mapping to the core model remains deterministic.

5. Consideration of Constraints

The last section showed the need for constraint declarations in the subtypes describing the mappings. Beyond that, the core model entities themselves may need to be provided with constraints. But whereas an attribute will be included in the core model only, if it has a correspondence in both original schemas, a constraint needs to be defined if it occurs in just one original schema. This is a consequence of the fact that all core model data shall be representable in both original schemas. A core object which violates a constraint of an original schema cannot be transformed into this representation.

The 7-schemas-model does not provide a formal derivation of constraints for the core model. It has been defined to describe how objects in the representation of one schema are related and can be transformed into objects in the representation of another schema. But it does not define how the other schema (e.g. the core model) has been created. The definition of the core model, i.e. the definition of its entities, their attributes and relationships, and the definition of the constraints is a process which relies on domain knowledge and human decisions and cannot be fully automated. Support of this process by expert systems is feasible. But the development of such tools would be a research area in its own right and is not the topic of the current investigation.

In the present approach, the origin of the core model constraints is indicated by comments in the EXPRESS code. Example 4 shows an extended description of the core model entity *cm_cell* of example 1 (also still simplified). An inverse attribute *owner* indicates that each *cm_cell* has to be used by exactly one library. This constraint has a correspondance in both original schemas, although it is expressed there in different ways. In EDIF, a domain rule is used to denote the cardinality constraint. Furthermore, a derived attribute (which relies on this domain rule) underlines the meaning of this direction of the relationship between *cell* and *library*. In CFI, an explicit attribute *Owner* is used. Two domain rules are needed to state that this attribute and the *Cells* attribute of the CFI *Lib* are inverse to each other. The unique rule of *cm_cell* requires that the cell name is unique within each library. The origin of this constraint is a domain rule in the CFI entity *Lib*.

```
ENTITY cm_cell;       -- Schema Core_Model
      name          : STRING;
      properties    : SET [0:?] OF cm_property;
      views         : SET [0:?] OF cm_connectivity_view;
INVERSE
      owner         : cm_library FOR cells;
-- EDIF: containment_constraint in cell
-- CFI: explicit attribute and constraints Lib.r2 and Cell.r2
UNIQUE
      name, owner;              -- CFI constraint Lib.r1
END_ENTITY;
```

Example 4

A detailed analysis of constraints in the information models of EDIF and CFI has been presented in [SchW94a]. A lot of these constraints can be replaced by specific structural EXPRESS mechanisms (INVERSE, UNIQUE, redeclaration). These structural equivalents are used for the core model as far as possible. For this purpose, the transformations described in [SchW94a] are applied in an informal way.

Although the derivation of the core model constraints is not visible in the 7-schemas-model, their existence is in no way arbitrary: when an object is transfered from one representation to another one, it has to obey the constraints of the target schema. Therefore when constraints are missing in the core model, there may exist objects whose transformation to an original schema representation would violate the constraints defined there. Such a counterexample would invalidate the given specification. On the other hand, when superfluous constraints are declared for the core model, the transformation capability is unnecessarily restricted which may impair an efficient use of the specification.

6. Conclusion

We have presented a formal frame, which allows to create a core model of given information models, which will describe the commonalities of the original models. In transformation schemas, subtypes will be defined which describe how to generate the core information from the original representations. Thereby a formal relationship is defined between the core model and the original schemas. A modification of the original schemas is not needed. If in addition the mappings from the core model to the original schemas will be defined, a transformation of data from one original representation into another one is possible via the core model. Thereby exactly those data can be mapped without loss, which are representable in the core model.

The concrete development of a core model for EDIF and CFI in [SchW94b] has proven that the approach is suitable for dealing with complex groups of entity types and their relationships. It allows to express even the case, that concepts of the original schemas only partially overlap. The inclusion of inheritance relationships (sub/supertypes) does not cause specific problems. This could not be shown here in detail. All constraints of the original models have to be included in the core model. Previous work [SchW94a] has given a classification of constraints and transformations of algorithmic constraints into a simpler structural form (UNIQUE and INVERSE clauses) for some frequently occuring constraint types.

A goal for our future research is the systematic analysis of the experiences of the concrete EDIF-CFI work to develop detailed rules for the unification work in the frame of the 7-schemas-model. The translation of the transformation schemas into EXPRESS-M [EXPM94] representations as a first step to the implementation of translators is another open topic.

For the conceptual description, EXPRESS-M is not the first choice. The relationship between entities of the core and original models is less strong in EXPRESS-M, which would impair the readability of the description and would require the definition of additional rules to restrict arbitrary mappings. The aspect of the transformation describing subtype as a view of the original entity would get lost. It is not obvious whether the instance control of EXPRESS-M fulfills our needs when mapping relationships between entities. We simply refer to the subtype describing the transformation of the target entity or (when necessary in mappings from the core model) make use of the EXPRESS entity constructor. The dependency of the EXPRESS-M instance control on the order of the mapping statements indicates, that EXPRESS-M is more suited for an implementation description than for a conceptual description of relationships between schemas. A difficulty for a translation into EXPRESS-M may be that EXPRESS-M does not support the complex entity construction operator.

The basic idea of our methodology is derived from concepts for schema integration in the field of heterogeneous databases. The concepts presented above will surely play also an important role for the definition of an integrated view, which provides all information of the underlying schemas. Whereas the core model represents a generalization of the schemas AtoC and BtoC (fig. 3), the integrated view can be regarded as a common specialization of both AtoC and BtoC: the information derived for the core model creation as well as the original information will be passed on to the entities of the integrated view. It will be a topic of future research, whether the definition of the integrated view can be described by means of EXPRESS in a similar way (e.g. using multiple inheriting subtypes) as it is possible for the core model.

References

[ACM91] ACM SIGMOD Records, Special Issue on "Semantic Issues in Multidatabase Systems". Vol. 20, No. 4, December 1991.

[BLN86] C. Batini, M. Lenzerini, S.B. Navathe: *A comparative analysis of methodologies for database schema integration.* ACM Computing Surveys, Vol. 18., No. 4, December 1986.

[CFI92] CAD Framework Initiative, Inc.: *Design Representation Programming Interface: Electrical Connectivity.* Version 1.0.0, Austin, TX, November 1992.

[EDIF93] *EDIF Electronic Design Interchange Format, Version 3 0 0 EIA-618.* Electronic Industries Association, 2001 Pennsylvania Avenue NW, Washington D.C. 20006, USA, June 1993.

[EXP92] ISO DIS 10303-11. Product Data Representation and Exchange - Part 11: *The EXPRESS Language Reference Manual.* TC184/SC4 Document N 151. August 1992.

[EXPM94] Ian Bailey (ed.): *EXPRESS-M Reference Manual.* ISO TC184/SC4/WG5 Document N 51. May 1994.

[IEEE91] IEEE Computer, Special Issue on Heterogeneous Distributed Database Systems, December 1991.

[SchW94a] G. Scholz, W. Wilkes: *Preliminary Report on Constraint Classification in Information Models.* ESPRIT project 8370 ESIP, Document No. ESIP/FUHAGEN/001.1, April 1994.

[SchW94b] G. Scholz, W. Wilkes: *Specification of a Core Model for Netlist.* ESPRIT project 8370 ESIP, Document No. ESIP/FUHAGEN/004.1, October 1994.

[SST92] M. H. Scholl, H.-J. Schek, M. Tresch: *Object Algebra and Views for Multi-Objectbases.* Proc. Int. Workshop on Distributed Object Management, Edmonton, Canada, August 1992.

PART FIVE

Communication and Distribution

12

Communication in a Distributed Environment

I. Videira, P. Veríssimo, H. Sarmento

IST/INESC
R. Alves Redol, 9, P-1000 Lisboa
Phone: +351 1 3100242 Fax: +351 1 525843
e-mail: imv@inesc.pt

Abstract

In this paper a new communication service is presented. A distributed architecture with communication agents in each tool and in each host is adopted. The environment is modelled as a set or a hierarchy of groups of components, and multicast communication is used inside and between those groups. This approach resulted in significants improvements in throughput, network traffic and fault-tolerance when compared to centralised approaches. For throughput, a comparison with ToolTalk demonstrates the better performance.

1 Introduction

Communication is a key issue in any kind of distributed system. System components may need to share data in order to perform a task. Management tools need to pass control information to other tools. When using replication, replicas need to communicate in order to maintain consistency. Users of a certain system may need to be able to share information.

Nowadays, specially with the migration from the "client-server" paradigm to the "object-oriented" paradigm, which is resulting in software systems being made up of a large number of objects spread over networks, the communication service is becoming even more important. The latter has to provide, not only the necessary interoperation, but also the ability to locate objects implementing the services required by other objects, and to activate them, if necessary. In addition, in certain systems, communication may need to be authenticated – objects may only grant access to some other specific and correctly identified objects.

This dependency on communication makes critical its reliability. Therefore, it is necessary to have dependable communication mechanisms, providing, namely, fault-tolerance and security authentication.

As we will see, in the CAD area, the communication services available do not present a satisfactory solution, mainly because they do not provide all the required mechanisms. Also, the current solutions in this area do not consider dependability.

It is the objective of our work to develop a new communication service offering the mechanisms needed in the Design Environment. Dependability is a major concern, and greatly influenced the adopted architecture.

Along the text we will refer to a *Design System* as a set of tools and data. A *Design Session* is a running instance of the Design System. A *Design Environment* is a set of Design Systems in an organisation.

In this paper we start by giving a short overview of currently available communication technologies. Then, we describe the communication mechanisms necessary in a Design Environment, which is seen as a particular case of a distributed environment. Finally, we present the Communication Service developed and the results achieved so far.

2 Communication technologies

In this section we present a brief overview of currently available communication technologies. In [5] a more detailed description can be found.

The conventional technologies for communication are message passing and remote procedure calls (RPC). Communication based on message passing mechanisms presents application interfaces at very different levels, most commonly at the OSI transport level. RPC technology hides the communication from the programmer, making a remote procedure call look like an ordinary local call.

RPC has been a very popular technique for building distributed systems, by the fact that it insulates the application from the networking details. However, this is done at the expense of flexibility. This flexibility can be achieved with the messaging systems, mainly with those providing low level interfaces. Since programming at low level is hard and requires a deep knowledge of the underlying protocols, some new messaging systems have been developed that offer a high level interface without loosing too much of the flexibility provided by the lower levels. The advantage of using messaging systems is that the application developer can build highly concurrent systems in a relatively easy way, having more control over the information exchange and flow than with the RPC services. However, for purely blocking calls, RPC is still easier to use than messaging systems.

The "object-oriented" paradigm is now becoming popular and people are finding out that it is adequate for building distributed systems. Management of distributed objects, including the communication mechanisms that support them, are now emerging technologies, and are mainly based on object brokers or traders that forward requests sent by an object to the most adequate providers. The Common Object Request Broker Architecture (CORBA) [1] from the Object Management Group (OMG) is the most known of these communication architectures.

The technologies seen so far permit point-to-point communication and invocation of procedures/methods on remote processes/objects, which may not be adequate in all the cases. For instance, in a group of loosely-coupled applications, where state information about a shared resource has to be passed to all applications, it is necessary to have mechanisms providing one-to-many communication (multicast). If the group can shrink and grow dynamically, it would be very difficult to implement this with the already described communication technologies. Another example where the conventional technologies are not adequate is in fault-tolerant systems. These systems normally use replication to achieve fault-tolerance, and, thus, they need other communication mechanisms like many-to-many between groups of component replicas.

In the areas of CAD and CASE the need for multicast communication led to the concept of "multicast message server" a centralised server through which all messages are routed. This server keeps a record of interests previously registered by tools and it delivers the messages received to the tools according to those interests. This communication architecture allows the interoperability of independently developed tools, as long as they use the same message set.

The concept of a multicast message server (in the context of CASE) was firstly introduced by Steven Reiss from Brown University and was implemented as part of the Field environment [3]. Several commercial products are now available using this concept, like ToolTalk from SunSoft, Broadcast Messaging Service from HP and Multicast Messaging Service from DEC. These products are used in CASE applications, respectively Sun SPARCworks, HP SoftBench and DEC Fuse. In the area of CAD the CAD Framework Initiative (CFI) is recommending an Inter-Tool Communication Service [2] also based on the concept of a multicast message server.

In the distributed systems world the concept of group orientation has been lately receiving great attention. Following this concept, distributed systems can be seen as sets or hierarchies of groups of cooperating tools. This kind of organisation raises some problems to its participants, like maintaining connectivity, managing group membership, doing reliable multi-participant communication, etc. In the recent years several solutions have appeared, such as group communication and group management.

3 Communication in a Design Environment

In a Design Environment (like in many other distributed environments) several communication mechanisms are required, with different qualities of service, to serve the specific needs of the components of the environment:

- A point-to-point mechanism with high throughput and/or low latency for closely cooperating tools (like a group of simulators).
- A communication mechanism with weaker requirements on performance, but simple to use, for loosely coupled tools. To make tools independent from each other, this mechanism should allow tools to exchange information without needing to know who owns the information or where it is located.
- A multicast mechanism to spread information about important events. It should be able to multicast events not only in the set of tools of the Design Session, but also in design teams, or in the whole environment, if required.
- A transactional mechanism for database access.

Currently, there is no single communication service fulfilling these requirements, which means that a tool developer needs to be familiar with several communication mechanisms in order to integrate his tools and to make them interoperate with the environment.

Furthermore, as the operability of the Design Environment is relying more and more on the communication between its components, it is necessary to have a dependable Communication Service. Therefore, this service must present continuous availability. Also, some authentication mechanisms need to be provided in order to protect data from being accessed by non authorised users. Until now these aspects were not considered in the CAD area.

The CFI Inter-tool Communication architecture is a first approach to the Communication Service required in a Design Environment, and, therefore, it is the basis for our work. We extended the used concepts in order to fulfill the requirements expressed above and to add dependability.

4 The Message Switcher

The conceptual architecture of the Communication Service is based on a component, designated by Message Switcher, which provides two distinct services: a Notification Service and a Request/Response Service (Fig. 1). Tool messages are delivered to the appropriate service by the Service Selection mechanism.

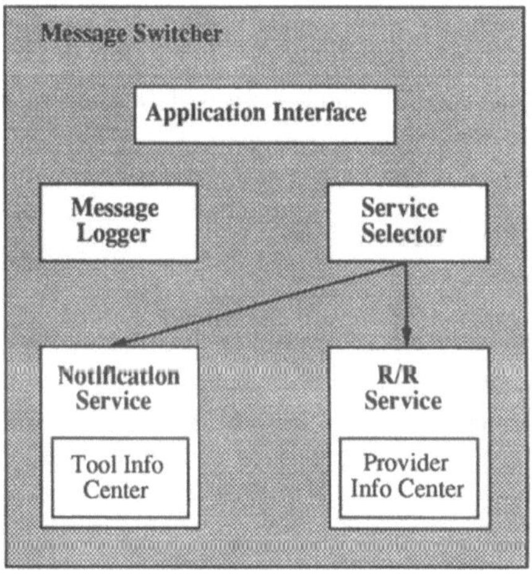

Figure 1: The Message Switcher conceptual architecture.

The Notification Service has basically the functionality of the CFI ITC Message Server, as it is able to forward messages to tools, based on the message contents and on the tools' interests. The Tool Information Center (TIC) is used to keep information about the tools, like status, communication end-points used and the list of message interests. The TIC can be loaded statically or dynamically. The objective of providing static tool registration is to enable the Message Switcher to start a tool, if necessary.

The Request/Response Service handles all the requests for data and routine/method execution. The Provider Information Center (PIC) maintains a record of the location of remotely callable routines or data (objects). This service is able to find the adequate provider for requests or to locate a specific provider, if requested to. The PIC can be statically or dynamically loaded. Static service (provider) registration allows the Message Switcher also to choose a provider that is not running, to serve a certain request.

The basic idea of the Message Switcher is to provide, through the same application interface, a notification service and a request/reply service, following, for the latter, the same philosophy of the former. This means that a tool willing to offer a certain service or to share some piece of data, uses the Message Switcher to advertise that. The information about what is provided and who provides it, is stored and used when a request is issued.

It is also possible to establish point-to-point connections using the Message Switcher. A tool willing to establish a direct link with another tool can ask that to the Message Switcher. Functions for dealing with point-to-point messaging are provided.

Being able to control all the services/data in the environment, the Message Switcher can also be used to control the access to those services/data using authentication mechanisms.

5 Implementation

For fault-tolerance and performance reasons mainly, we decided to adopt a distributed solution in the implementation of the Message Switcher. The service is distributed by two types of entities: an agent in every host of the Design Environment – the Host Agent (MSHost), and an agent linked with each tool – the Tool Agent (MSTool) (Fig. 2).

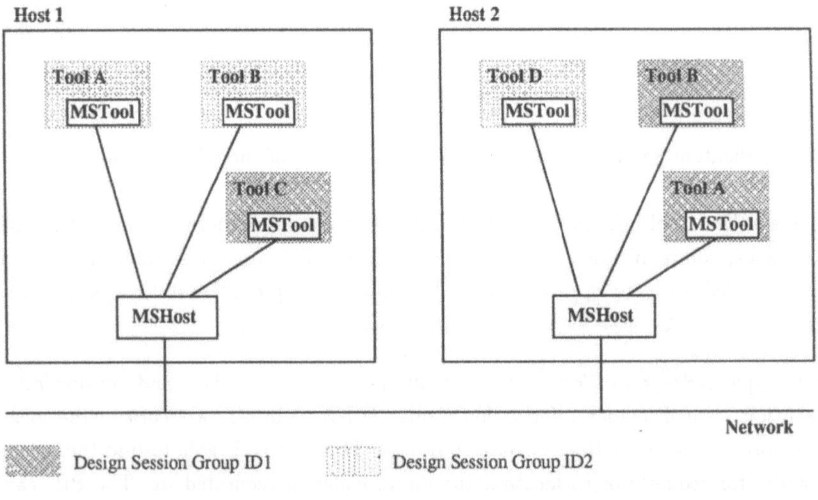

Figure 2: The Message Switcher agents.

Communication is done inside multicast groups. Several different groups are possible, like the Design Session Group, the Design Team Group, or the Design Environment Group. Messages from a tool in a Design Session under a certain user ID, can be multicast to any group to which the user belongs. Assignment of users to groups (others than the Design Session Group) is done by the system administrator or the project manager.

Tools register in the Host Agent using functions of the Tool Agent. The Tool Agent is responsible for sending private messages directly to some specific tool, or to a multicast group. Notifications or requests that do not specify a provider are always multicast to the Design Session Group, and will be received by every Host Agent having tools registered in that Design Session. A response to a request is also multicast to the Design Session Group and can, therefore, be seen by the interested tools.

A Host Agent is responsible for forwarding the messages received within the group to the interested tools in its host and to find adequate providers for requests. Each Host Agent keeps information only about tools running in its host (and about tools that registered statically).

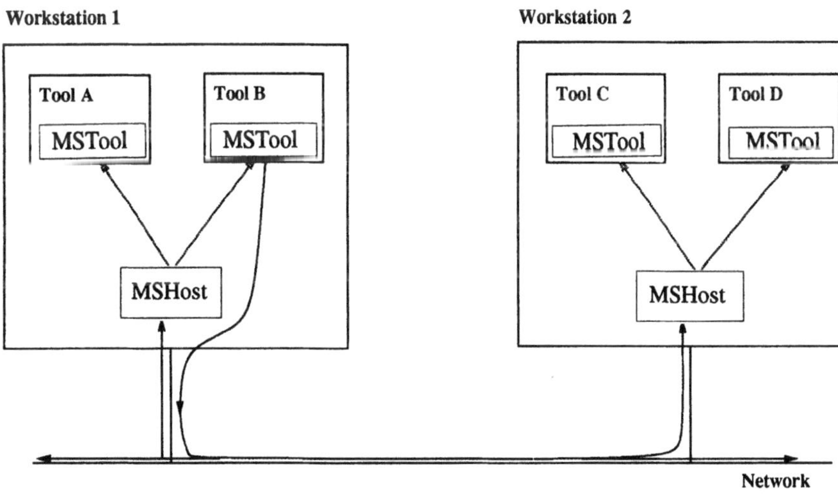

Figure 3: Notifications path.

Figure 3 shows the path followed by a notification: 1) it is sent by the Tool Agent of Tool B to the multicast address of the Design Session Group; 2) each Host Agent that serves tools in that group receives that notification; 3) each Host Agent consults its TIC to verify if any of the tools it serves is interested in that notification; 4) each Host Agent forwards the notification to the tools interested. It is important to highlight that message forwarding is done in parallel in both Host Agents.

Inter-Session communication can be easily done by explicitly multicasting messages to specific Design Session Groups, provided that the tool has permission to do it. Otherwise, following the principle of sending "to whom might be interested", tools can register interest in receiving specific messages from other Sessions. The Host Agent will forward them the inter-session messages that match with any of their interests.

This distributed implementation of the Communication Service is inherently fault-tolerant to network failures, because there is an agent of the service in every host that can work isolated. Also, host failures are not a problem, since a Host Agent only serves tools in its host. Therefore, it is only necessary to mask process failures of the Host Agent, which is very easily achieved.

Multicast communication is implemented with IP Multicasting through the BSD socket interface. As the delivery of multicast datagrams is not reliable (as in ordinary IP datagrams) it is necessary to implement a reliable protocol on top of it. In a LAN (which is currently the environment of a Design Session) the loss or duplication of packets is not very frequent. Therefore, the introduction of reliability will have a minor impact on performance, because there will be few retransmissions.

6 Performance evaluation

Some tests were done, in order to compare the performance of our approach with the available results for ToolTalk [4]. These tests only evaluate the performance of the Notification Service for two reasons: first, at the time the Request/Response Service was not yet entirely finished and second, there are no published figures for ToolTalk performance in serving requests.

We adopted the same test scenario used for obtaining the ToolTalk results:

- Tests were done on a networked SUN IPX workstation
- In each test a tool sends 5000 notifications
- Each notification has a size of 200 bytes
- Each tool receiving a message has two interests registered
- The tool supplied callback routine, to be called at message arrival, performs no operation

- Notifications are sent within the scope of a Design Session

In Table 1 it is shown the elapsed time between sending the first notification and receiving the last one in each tool, as a function of the number of tools receiving the messages. Figures are shown for the Message Switcher and for ToolTalk.

		Sender	Receiver 1	Receiver 2	Receiver 3	Receiver 4
1 tool	**Mes. Switcher**	**19 s**	**19 s**			
	ToolTalk	97 s	97 s			
2 tools	**Mes. Switcher**	**24 s**	**24 s**	**25 s**		
	ToolTalk	153 s	153 s	154 s		
3 tools	**Mes. Switcher**	**28 s**	**28 s**	**29 s**	**30s**	
	ToolTalk	206 s	206 s	208 s	207 s	
4 tools	**Mes. Switcher**	**29 s**	**29 s**	**30 s**	**31s**	**34s**
	ToolTalk	266 s – *Average value for all columns*				

Table 1: Comparison of the Message Switcher and ToolTalk performance.

Another test was conducted, in order to evaluate the performance when tools are running in different hosts. For this test we had two tools running in different hosts, where one is the sender and both receive the burst of notifications. We verified that, as expected, the obtained throughput is exactly the same as if only one receiving tool exists(5000 messages in 19 s). Delivery in done locally in each host, not affecting delivery in other hosts.

7 Conclusions

In this paper a new Communication Service using a distributed architecture that resulted in considerable advantages over the current centralised solutions, is presented. Those advantages can be summarised as follows:

- A great reduction in network traffic: each multicast message is only sent once to the net and arrives to every Host Agent in the Design Session.
- A great improvement on performance: due to network traffic reduction and to the fact that message filtering and forwarding is a distributed task, performed locally in each Host Agent.
- Achievement of fault-tolerance to host and network failures.

Performance evaluation experiments could confirm the expected improvement in throughput of our approach when compared with a centralised one. The other two referred advantages are obvious (do not need confirmation).

Currently, the Notification Service and the Request/Response Service implementations are finished. We are implementing the CFI ITC interface specification on top of the Message Switcher. This will allow tools using CFI ITC to also use our component, and, namely, it will allow the integration of the Message Switcher in the JESSI-COMMON-FRAME framework.

As future work, we intend to further develop the Message Switcher. Several new features that will provide better support for group communication will be added, like support for group definition and management. Furthermore, other features that will, in geral, improve the service will also be added, such as the integration of a load balancer to the request/reply mechanism, configuration mechanisms, message priorities and support for communication at the user level.

Acknowledgments

This work has been partially funded by the European Commission Esprit Research and Development Programme in the sector of Microelectronics under Project JESSI-COMMON-FRAME (#7364) and by Junta Nacional de Investigação Científica e Tecnológica under Project STRDA/C/TIT/78/92.

References

[1] *The Common Object Request Broker: Architecture and Specification*, Object Management Group and X/Open, December 1991.

[2] *Inter-Tool Communication Architecture, Version 0.9.1-012292*, CAD Framework Initiative, January 1992.

[3] Steven Reiss, "Connecting tools using message passing in the Field environment", *IEEE Software*, July 1990.

[4] Results published in CFI mailing list "cfi-itc@cfi.org", June 1993, and USENET newsgroup "alt.sys.tooltalk" FAQ, 1993.

[5] Idalina Videira, *Report on the analysis of various communication services*, ESPRIT 7364 Jessi-Common-Frame, Deliverable D1.ITC-1, November 1992.

13

Basic Requirements for an Efficient Inter-Framework-Communication*

Arno Kunzmann **Ralf Seepold**

Forschungszentrum Informatik, FZI
Systementwurf in der Mikroelektronik (Prof. W. Rosenstiel)
Haid-und-Neu-Straße 10-14, D-76131 Karlsruhe, Germany
E-mail: kunzmann/seepold@fzi.de

Abstract

For the design of complex electronic CAD systems, several frameworks have been proposed and implemented offering a different set of basic services, and therefore, supporting specific and dedicated design activities. In this paper the results of a case-study are summarized where two almost complementary frameworks have been coupled in order combine their advantages: (1) the JESSI-COMMON-Framework with special emphasis on design management, and (2) FMCAD (representing a widespread design environment of the electronic CAD domain) that is tuned on close tool integration and efficient data handling. As a result of this case-study the basic requirements for an efficient inter-framework-communication are described to indicate some trends for future framework developments.

1 Introduction

Within the electronic CAD domain there exist several frameworks with a different set of serivces resulting in a specific support of dedicated design activities. For many applications a combination of these isolated environments could improve design efficiency. Therefore, the current work focuses on the coupling of two different, almost complementary frameworks: the JESSI-COMMON-Framework (JCF) release 3.0‡ [17] and a widespread design environment of the electronic CAD world that will be called FMCAD in this paper.

As mentioned above, each framework has its own strengths. The JCF framework is a very powerful environment for supporting design management, configuration management and concurrent engineering. Therefore, JCF does not focus on a certain domain, but supports a set of services which are required in many different domains. In contrast to that, the function of FMCAD is concentrated on close tool integration and efficient interaction for specific prob-

* This research is supported by the ESPRIT project 7364.

‡ SNI provides a commercial version of JCF under the name SIFRAME.

lem-oriented electronic CAD environments. In this case, the guidance and support of a designer are only rudimentarily provided.

The ultimate goal is to combine the main advantages of both approaches to form an extendable and powerful environment for the development of electronic systems and to derive fundamental requirements for an efficient inter-framework-communication. Subsequently, the main differences of the isolated systems and the aiming advantages of a combined approach are discussed.

2 Presentation of the JESSI-COMMON-Framework and FMCAD

In this chapter the JESSI-COMMON-Framework (JCF) and the FMCAD framework are presented to give a brief overview of the main structuring (e.g. architecture). Additional information can be found in [6], [8], [9] and [16].

2.1 JESSI-COMMON-Framework

JCF release 3.0 was chosen for the for framework coupling case-study because this release was available at the beginning of this work. Essential parts of the JCF information architecture can be found in [9], the main basic concepts are summarized in this subsection.

JCF applies a strict distinction between *resources*, which can be used for different projects, and actual *project data*. At the *resources definition*, the JCF administrator defines users and teams that are allowed to use the framework. Furthermore, design flows [11] are precisely defined. For this purpose it is described which tools are part of the design flow, how these tools can be used, and which data are passed between the tools. Note that these resource data are stored as meta data in the framework, and are completely under control of the framework. This is realized in contrast to FMCAD which does not manage any resource data.

The kernel of the JCF *project organization* is formed by a product structure that is modelled via a *Cell* hierarchy (more precise: via *ComposedOf* relations between *Cell Versions*). Each *Cell Version* has a flow and a team attached to it. The philosophy is that only members of the team are allowed to work on such a *Cell Version*, and that they must follow the attached flow. A particular *Cell* can have several *Cell Versions*, with possibly different *teams* or *flows* attached.

When a user wants to work on a *Cell Version*, he must first *reserve* it in his *Workspace*. It is then isolated for his own use, and no other user can do any work with it. As soon as the work on the *Cell Version* is finished, the user *publishes* it. The *Cell Version* is then removed from the *Workspace*, and can be used by the other users of the *team*.

JCF stores all its data in the underlying object-oriented OMS database. As soon as data are needed, they are checked-out from OMS, and when work is finished, data are checked-in again. The fact that a real database is used for data storage is a significant difference to the FMCAD approach, where data are just stored in the UNIX file system. For more information, we refer to the JCF documentation (e.g. [6] and [12]).

2.2 FMCAD

FMCAD stores design data in libraries. A library consists of a directory in the Unix file system and a library catalog file. The catalog file stores the logical (meta) data objects. The design data files are stored in the Unix directory. The basic logical data objects in a FMCAD library are *cells*, *views*, *cellviews*, *cellview versions* and *configurations*.

A *cell* is the basic design object. It forms an individual building block of a chip or system. It is a logical, rather than a physical design object. Each *cell* has one or more *cellviews*.
A *view* is one type of representation, such as schematic or layout. A *view* has a name and is of one specific *viewtype*. The *viewtype* associates the *view* with a certain FMCAD tool, for example the layout editor. A *view* is also a logical design object. A *cellview* is a virtual data file created in association with a *cell* and a *view*. It is a logical, rather than a physical design object. A *cellview version* is the data file of a *cellview* at a particular time. *Versions* are created by check-out/check-in operations. A *cellview* can have more than one *version*. The *version* models the link to the design file in the file system. The checked-out *version* is the *version* which a user is working on. Only one *version* of a *cellview* can be checked out at a time. A *configuration* is a collection of *cellview versions* that are related. For each *cellview* at maximum one *version* can be part of the *configuration*.

2.3 Basic Results

The combined JCF-FMCAD framework approach should sum up the advantages of both approaches to form a new and more powerful environment for chip design. Of course, it is not possible to combine only advantages without reducing the high flexibility of FMCAD, but as described in this paper an efficient design support can be enabled. A very essential task of the framework coupling is the mapping of the different data models. In order to support the individual characteristics of the frameworks, the master-slave approach was selected. Based on the very powerful management features of JCF, it was decided that JCF is the master framework. A more detailed description of the data mapping is given in [9].

In the following, the main achievements including the basic problems of the combined JCF-FMCAD framework approach are evaluated and summarized. A short overview is provided in Table 1 and a detailed description of the results is stated by the following bullet points:

- **Multi-User Design and Concurrency Control**

  One achievement was to use the multi-user capabilities of JCF 3.0 also in the JCF-FMCAD environment. Based on the JCF *workspace* concept, individual design elements can be reserved and isolated from the access of other users. While in FMCAD, parallel work on different versions of the same *cellview* is not possible, the JCF-FMCAD environment provides this feature.

- **Design Management and Data Consistency**

  Compared with FMCAD, the combined framework provides a more flexible design management system. For instance, JCF-FMCAD offers the versioning of both *cells* and *design objects*. In addition, the representation of hierarchy information by meta data allows a more powerful consistency check.

- **Handling of Design Hierarchies**

  This was one of the most difficult tasks during the JCF-FMCAD integration. Compared with FMCAD some flexibility in handling design hierarchies was lost, since the existing JCF-FMCAD prototype requires that all hierarchical manipulations must be done manually via the JCF desktop. Another limitation is caused by non-isomorphic hierarchies that are not yet supported by JCF release 3.0 but will be supported by future releases.

- **User-interface**

  In the JCF-FMCAD prototype system, the designer has to work with both the FMCAD and JCF user-interface. In spite the fact that JCF is conform with the major user-interface standards (using X-Windows and Motif), the user has to cope with an additional user-interface.

- **Flow Management and Derivation Relations**

  In contrast to standard FMCAD, JCF-FMCAD provides flow management capabilities, but in a slightly restricted way, due to the available handling of non-isomorhic hierarchies in the current JCF release 3.0. Especially to support the archiving and retrieving of design data, derivation relations and *what-belongs-to-what* information are very essential features that are supported by the JCF-FMCAD framework system.

- **Performance**

  This aspect is of lower importance since the main aspect of this inter-framework coupling is functionality. The performance of meta data operations in the JCF-FMCAD environment is sufficiently high.

Criteria	JCF 3.0	FMCAD	JCF-FMCAD
Multi-User Design	+	-	+
Concurrency Control	+	o	+
Design Management	+	o	+
Data Consistency	+	-	+
Design Hierarchies	-	+	o
User Interface	+	+	o
Design Flow Management	+	-	+
Derivation Relation	+	o	+
Performance	o	+	o

+ ≡ good support
o ≡ medium support
– ≡ weak support

Table 1: Criteria Table

3 Requirements for an Inter-Framework Communication

Based on the performed case-study, a major task of future developments for inter-framework-communication is to close the gap between isolated and stand-alone frameworks. Existing framework isles have been claimed for industrial standards and are very important for long-term company decisions. The following requirements are derived from the described experiences that are based on a fundamental mid-term study and several consultations with major chip developers. As a result, Table 2 summarizes the main requirements that can be regarded as general requirements and should be taken into account for future developments of open frameworks.

Requirements	Priority	Advantage
efficient data access	high	enhanced performance
support of standards (e.g. CFI)	high	easy integration and customization
support of user-interface customization	medium	provide a common look & feel
modularity	high	adequate and flexible customization
support of project management	high	reduced costs by efficient use of resources
design management functionality	high	flexible modelling
shared libraries	medium	data consistency
support of integration	high	fast and tight tool integration
support of ITC	high	efficient use of standard capabilities

Table 2: Requirements for Inter-Framework Communication

The coupling of different frameworks requires the mapping of different data models. Furthermore, the exchange of meta and design data is necessary to communicate between the framework sites. An efficient data access influences both the performance and acceptance of the coupled system. The priority of this requirement is very high because the end user has to be convinced to use the system. Limitations exist in case of closed interfaces or import/export functions that are based on file-formats. Future framework releases have to take into account that the mentioned aspect is of high priority.

The coupling of very different systems is hardly manageable if existing standards are ignored or missing. CFI standards should be supported as soon as they are available to guarantee easy and efficient integration of environments.

Due to the fact that each user prefers a customized environment to enhanced his own work profile, a modern user-interface should be very flexible and easy to configure. Of course, this requirement is of medium importance.

Regarding the different existing environments in IC companies, each major company mixes in-house developed tools with commercial tools or frameworks. In many cases, the optimal solution is not to integrate the whole tool set, but some specific modules, since a full integration requires too much time and effort. Furthermore, a module concept provides interfaces to customize and configure a framework that fulfils exactly the requirements of the company. If standards (e.g. CFI) are missing other preliminary interfaces have to be offered. The final transfer should be supported by migration tools.

The deep integration of design tools is a very complex task. But the management of a complex project is much more important than a deep integration. Wrong decisions on this level will waste time and money and may cause the stop of a project. Efficient project management facilities which support the project manager through all phases of the development are completely missing. Additionally, new concepts have to be developed to close the gap between existing systems in the areas of Operation Research and the existing ECAD frameworks.

Several different levels of designer support can be distinguished (e.g. knowledge based systems or design flows). The acceptance of the system by the designer, who is a specialist in using a subset of the design tools, will be influenced by the flexibility of the design management. The design management must be flexible to avoid strictly forced design flows were possible, and it must be applicable to control the project state and ensure a specific quality level. On the one hand a bypassing of flows should be impossible, and on the other hand the flow should offer more flexibility than pure logical AND connections.

The use of libraries is a central concept in chip developments. It is very difficult to keep the library consistent and continue work during the design while changes are taking place. Concurrent engineering should be offered for the designer's work as well as for system administrators to ensure consistency at all levels.

It is important to notice that there will be different levels of integration for external tools. Only a minor set will be deeply integrated and the majority of tools can be adapted by a loosely coupling. Nevertheless, integration facilities must be available, very efficient, supported by standard tools and conform to existing standards. Many commercial systems support this facility by a powerful extension language.

Concerning the modularity of software and software products, inter-tool-communication (ITC) is a future-oriented feature to offer very flexible and powerful mechanisms for tool coupling. This coupling is necessary for tools which communicate among others inside the same framework and also between different frameworks. This service should be offered as a basic tool for integration and communication.

4 Conclusion and Future Work

With the described JCF-FMCAD framework prototype it could be demonstrated that a combination of frameworks with different basic architectures and individual design services can provide enhanced functionality. The study provides basic results which can be used to enhance the state-of-the-art in inter-framework-communication and push future framework development.

In some cases, the coupling offers not the advantages of both isolated frameworks but this fact is mainly influenced by the existing software releases, i.e., limitations in available interfaces. The coupling of frameworks depends on the experience of the integrator and is not supported by existing tools, therefore, the data model mapping remains a key aspect for successful integration.

The new JCF release 4.0 will offer more open interfaces and enhanced functionality. The integration of tools and the coupling of other framework isles should be easier and more efficient than in the existing system. A planned completely new integration and coupling will show, if the demanded requirements are fulfilled. Furthermore, the level of support for inter-framework-communication (e.g. ‚work in progress' or ‚archive') has to be defined with respect to technical and praxis-oriented requirements for the next generation of design environments. This objective can be regarded as the next step towards a new prototype.

Acknowledgements

We would like to thank Hans Kok, Kees Lepoeter (Philips Research, Eindhoven) and Thomas Harriehausen (Siemens HL, Munich) for their contributions, support during the study and the fruitful discussions which enabled this publication.

References

[1] W. Allen, D. Rosenthal and K. Fiduk, *'The MCC CAD Framework Methodology Management System'*, Proc. 28th DAC, pp.694-698, 1991.

[2] K.O. ten Bosch et al., *'Design Flow Management in the Nelsis CAD Framework'*, Proc. of 28th DAC, pp. 711-716, 1991.

[3] F. Bretschneider, Ch. Kopf and M. Zolg, *'Process Control for Large Complex Systems'*, Winter Simulation Conference, 1991.

[4] Cadence Design Framework II Reference; Cadence release 4.2.2 documentation.

[5] Framework Criteria Definitions, Authors: SP4 Team, Editor: H. Dorfmeister, Internal Report, ESPRIT Project 7364, JCF/BOSCH/001-02/25-September-1992, 1992.

[6] JCF 3.0 System Overview; Documentation JCF release 3.0.

[7] Y. Kashai, *'Flow - A Concurrent Methodology Manager'*, Proc. EDAC, pp. 20-24, Paris, 1992.

[8] H. Kok, K. Lepoeter and R. Seepold, *'Report on the JCF-Cadence encapsulation project'*, ESPRIT Project 7364, Internal Report, JCF/FZI-Philips/032-1/18-May-94, 1994.

[9] A. Kunzmann and R. Seepold, *'Enhanced Functionality by Coupling the JESSI-COMMON-Framework with an ECAD framework'*, Proc. EDAC, Paris, 1995.

[10] E. Kwee-Christoph, B. Eschermann, O. Haberl, R. Kumar and A. Kunzmann, *'The CADEC VLSI Design Support Methodology'*, CompEuro, The Hague, The Netherlands, 1992.

[11] D. Liebisch and A. Jain, *'JESSI COMMON FRAMEWORK Design Management -- The means to Configuration and Execution of the Design Process'*, Proc. Euro-DAC, pp. 552-557, Hamburg, 1992.

[12] A. Meckenstock and D. Zimmer, *'Concept and Architecture of a Distributed Object-Oriented Database Kernel'*, Cadlab Report 13/92, CADLAB Paderborn.

[13] S. Parikh, D. Sarnoff, M.L. Bushnell et al., *'Distributed Computing, Automatic Design, and Error Recovery in the ULYSSES II Framework'*, Proc. EDAC, pp. 610-617, Paris, 1994.

[14] T. Quellenberg, J. Schubert and G. v. Bültzingsloewen, *'FLOPPY - A Flow Management for On-line Processes in Production Systems'*, submitted to IMSE European Workshop on Integrated Manufacturing Systems Engineering, Grenoble, 1994.

[15] F. J. Rammig and B. Steinmüller, *'Frameworks und Entwurfsumgebungen'*, Informatik-Spektrum (1992) 15, pp.33-43, 1992.

[16] J. Schubert, R. Seepold, A. Kunzmann and U. Weinmann, *'Modelling a FPGA Design Flow in the JESSI-COMMON-Framework'*, The 10th ISPE/IFAC International Conference on CAD/CAM, Robotics and Factories of the Future CARs&FOF'94, pp. 680-685, Ottawa, August 1994.

[17] B. Steinmüller et al., *'JCF-Start-up Phase Final Version'*, Internal Report, ESPRIT Project 7364, JCF/CADLAB/053-02/01-June-92, 1992.

[18] M. Straube, W. Wilkes and G. Schlageter, *'HANDICAP - A System for Design Consulting'*, Proc. EDAC, pp. 600-604, Paris, 1994.

14

Resource-Oriented Load Distribution in a Framework Environment *

Jürgen Schubert **Arno Kunzmann**

Forschungszentrum Informatik, FZI
Systementwurf in der Mikroelektronik (Prof. W. Rosenstiel)
Haid-und-Neu-Straße 10-14, D-76131 Karlsruhe, Germany
E-mail: schubert/kunzmann@fzi.de

Abstract

In this paper the framework-based load distribution system FLODIS is presented. In order to reduce the overall design time, FLODIS provides a user-transparent distribution of design activities over a heterogeneous network of workstations. The distribution algorithm is mainly controlled by estimations about the required activity-specific resources that can be derived by user-defined data, recorded execution profiles and methodology information of the framework. Experimental data with several design flows and different design activities show that by the proposed distribution system the overall execution time can be distinctly reduced. Compared with conventional distribution systems an average reduction up to over 60% can be observed.

1 Introduction

The development of complex systems requires a highly efficient cooperation of design teams to reduce the overall design time. A key objective is the support of concurrent engineering. This necessarily implies that a simultaneous and coordinated execution of a large number of different design tools is required. Depending on the selected parameters and the input/output data, each tool can perform several activities. Given a network of several workstations, the execution time for each activity is heavily influenced by the selected workstation. In order to reduce both the individual and the overall response time, several load distribution systems have been developed. Since the management of complex design tasks is typically embedded in a specific design framework, the efficiency of load distribution systems can be significantly improved by using framework knowledge.

*This work is supported by the ESPRIT project JESSI-COMMON-Frame (7364).

In this paper, the concept and implementation of the load distribution system FLODIS (Framework based LOad DIstribution System) will be discussed. Experimental results show that compared with conventional load distribution approaches, FLODIS distinctly reduces the overall response time in average up to over 60 %.

2 The FLODIS Approach: Motivation and Concepts

One general problem of most load distribution systems is the missing knowledge about the upcoming load. Hereto, static approaches use statistics, for instance, fixed job arrival rates. The distribution is performed without measuring the load of the available workstations, assuming an average external load [TT85]. In contrast, dynamic distribution algorithms consider the current system load, in order to select the workstations with least load.

Embedded in framework environments, several approaches were reported: The MCC CAD framework selects an appropriate workstation based on the (static) Round-Robin algorithm [ARF90]. FLOW is a representative approach for dynamic load distribution [Kas92]. PAPYRUS is based on the operating system SPRITE and, therefore, also process migration is supported [CK93]. In the ULYSSES II framework, resources are dynamically allocated [Bus94]. Here, the best suited workstation is selected by calculating the processing power and work load. A common feature of all these distribution strategies is that they are only based on the number of activities. But obviously, the necessary resources (e.g., CPU-time, I/O-time, main memory) are essential to compute efficient allocations. Therefore, FLODIS also takes resources into account, but since exact values are only available in very special cases, estimations about the required resources will be used. In a framework environment precise predictions about activity specific resources can be gathered by different sources:

a) The tool integrator often knows which complexity dimension (CD) value (e.g., number of gates, transistors, primary inputs) are important for computing the activity specific resources. Additionally, detailed resource requirements, e.g., how the number of gates influences the memory and CPU-time consumption may be known. For several activities like simulation and test pattern generation, an average complexity of the basic algorithm is available and the corresponding calculation formula can be used for load predictions.

b) With every activity execution, the used resources are recorded, always related to the CD values. In FLODIS a linear regression is used to forecast CPU-time, I/O-time and memory requirements [DS66].

c) Derived from the methodology management, the information of repeatedly executed activities within a subproject can be used. In this case the CD values are expected to be similar because the same specification is used for the part which has to be designed. The mean value of the previously used resources is sufficiently accurate.

d) For executions within the same design flows, where predefined activity sequences have to be followed, the CD value of the last execution is chosen, in general providing an even more accurate value. This is a typical case because a design process often is iterative which includes the re-execution of activities with slightly modified input until the specification is fulfilled.

To establish the required communication between framework and load distribution system, the framework sends activities to FLODIS together with the information about the identification of the design flow and the subproject from which the activity is started. In FLODIS the resource estimation algorithm in Figure 1 is used for characterization with respect to the relative CPU and I/O usage of the activities to distinguish between CPU and I/O intensive jobs.

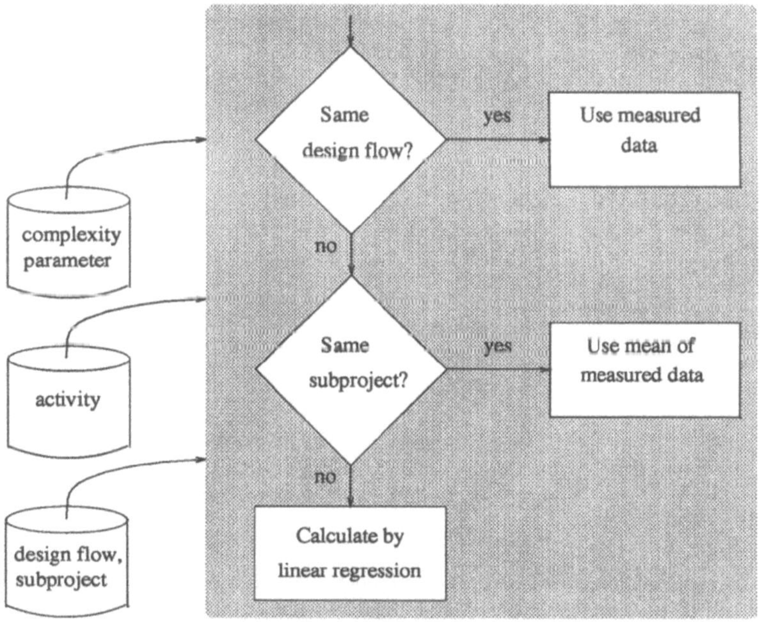

Figure 1: Activity resource estimation algorithm

Since an execution of jobs with the same characteristics on the same workstation results in relatively high increases of the response times, in FLODIS an allocation strategy is implemented that prefers the combination of tools with different characteristics. Given a workstation specific maximum CPU and I/O load, activities can be allocated as long as this limit is not exceeded. To take also workstations with different computing power into account, transfer factors for adapting CPU and I/O requirements are calculated. In order to optimize the use of powerful workstations, the distribution algorithm is based on a list of available workstations, descending ordered according to their computing power. Always starting at the top of this list, activities are allocated if the available resources fulfil the activity requirements. In this case, the workstation resources are decreased according to the resource predictions for the allocated activity, otherwise the next workstation is scanned.

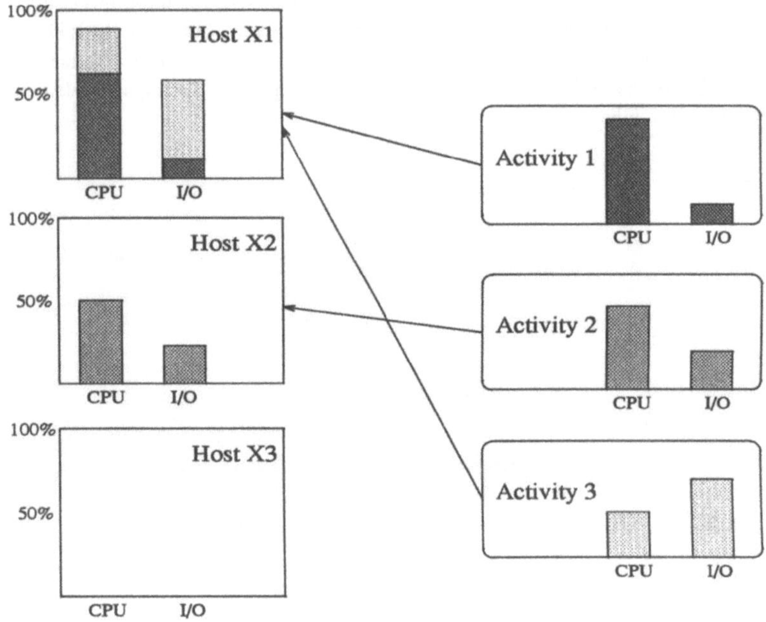

Figure 2: Activity allocation algorithm

Figure 2 presents an example for the allocation of three activities that have to be executed. The available three hosts are ordered according to their computing power. Both activities, activity 1 and activity 2, are CPU intensive jobs. The allocation algorithm selects the more powerful host (Host X1) for execution of activity 1 and the available resources of this workstation are decreased by the estimated activity requirements. The second activity needs more CPU resources than host X1 can provide, therefore, the less powerful host (Host X2) is selected. The third activity needs a lot of I/O resources. The allocation algorithm tries to allocate the activity at the most powerful host X1. As both, the CPU requirements and the I/O requirements, do not exceed the predefined limit, host X1 is selected. This guarantees on one hand that the activities don't have to wait for resources which are used by another activity, and on the other hand, the activities are executed on the most powerful workstation which is available. The minimum execution time of the activities is provided in spite of the fact that the host X3 is not used.

The presented algorithm does not take the real load of the workstations into account. It is assumed that in a framework design environment all the activities on the workstations are controlled by the framework. Of course, this assumption does not hold, if the designer is allowed to start a process which is not known by the framework. Therefore, FLODIS offers the possibility to refuse an execution if an overload of a workstation has to be avoided. FLODIS checks the selected host before an activity is sent for execution. If the host does not provide the required resources because non framework processes are running, the activity is sent back to the allocation algorithm for a re-selection of another host. Additionally, the specific host is marked as 'overloaded'. The execution of activities is not delayed because only the selected host has to be checked and not the workstation environment. Of course, a delay occurs in case of an overloaded host, but this is assumed to be an exception.

3 Experimental Results

The reported results are computed by a prototype implementation of the FLODIS system based on the JESSI-COMMON-Framework (JCF) [LJ92]. At present FLODIS is implemented as a domain neutral tool that is loosely coupled with the JCF framework, i.e. all necessary data for the load distribution are exchanged via the JCF Tool Management System. The experimental analyses are based on three different types of design flows. As indicated in Figure 3, sub-flows can be executed in parallel. The degree of parallelism characterizes the three basic flow types, where (n,m) is set to (2,1), (3,3) and (6,6). Obviously, in the last case 24 activities have to be executed.

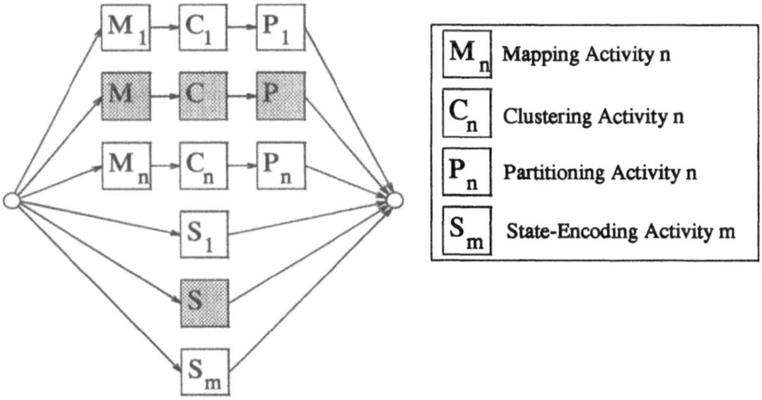

Figure 3: Principle structure of the evaluated design flows

All evaluated flows are typically used for the design of field-programmable gate-arrays (FPGAs), and consist of four basic activities for mapping, clustering, partitioning and state encoding. To receive practice-oriented measurements, all flows were executed with up to 12 different control parameters and input data, e.g. circuit sizes. For the subsequently given response times, a network of six SUN workstations with different computing power is assumed (1xSS20, 1xSS10, 2xSS2, 2xSS1). In Table 1 and Table 2, the results for two conventional approaches (static: Round Robin and dynamic: LB [KS94]) and FLODIS are listed, based on three different settings of n and m.

	n=2,m=1			n=3, m=3					
	Parameter Set			Parameter Set					
	1	2	3	1	2	3	4	5	6
Round Robin	243	88	126	333	171	462	87	212	589
Dynamic (LB)	243	106	103	261	135	114	127	200	302
FLODIS	98	49	72	211	74	97	65	88	164
Savings RR [%]	59,7	44,3	42,9	36,6	56,7	79,0	25,3	58,5	72,2
Savings LB [%]	59,7	53,8	30,1	19,2	45,2	14,9	48,8	56,0	45,7

Table 1: Response times for two different flow types with 3 and 6 parameter sets [sec.]

	n=6,m=6											
	Parameter Set											
	1	2	3	4	5	6	7	8	9	10	11	12
Round Robin	539	548	202	67	130	69	132	159	340	100	221	290
Dynamic (LB)	226	310	154	53	133	64	119	152	317	118	173	249
FLODIS	176	256	114	49	123	50	121	180	260	81	128	186
Savings RR [%]	67,3	53,3	43,6	26,9	5,4	27,5	8,3	-13,2	23,5	19,0	42,1	35,9
Savings LB [%]	22,1	17,4	26,0	7,5	7,5	21,9	-1,7	-18,4	18,0	31,4	26,0	25,3

Table 2: Response times for the third flow type with 12 different parameter sets [sec.]

The given response times show the impact of the selected parameters. Besides these times values, for each parameter a direct comparison between the Round-Robin, LB and FLODIS approach is given by the last two rows. With only a very few exceptions, the ranking between these three systems is obvious: the dynamic LB has slight advantages over the static Round Robin approach, whilst the proposed new method requires the lowest response time in average. This result is also stated by Table 3, where the individual response times are summed up. Compared with Round-Robin, the relative response time savings of FLODIS are between 38 % and 62 %, for LB the achieved reductions range between 16 % and 51 %.

	n=2 m=1	n=3 m=3	n=6 m=6
Round Robin	459	1857	2802
Dynamic (LB)	454	1142	2074
FLODIS	219	701	1728
Savings RR [%]	52,3	62,3	38,3
Savings LB [%]	51,8	38,6	16,7

Table 3: Comparison of the overall response times [sec.]

The most significant savings can be achieved if a small number of activities is executed in the workstation net (n=2/m=1). This is the typical case in design environments. However, FLODIS also provides better results if the number of activities is higher than the available number of workstations.

4 Conclusions and Future Work

Short time-to-market is one main objective in the development of new products. Especially the design of complex systems where extensive CPU usage can be foreseen, an optimized usage of the available workstations can provide distinct response time savings. Within a framework environment FLODIS takes advantage of the available information about activities, project hierarchy and predefined design flows. Controlled by the current system load and the design activity profiles that allow the prediction of the expected CPU and I/O behaviour, an optimized allocation of activities to workstations can be performed. In contrast to conventional static load distribution systems, FLODIS allocates systematically several activities to one workstation as long as a load limit is not exceeded. As underlined by the experimental results the proposed distribution strategy shows distinct savings of the necessary overall response time up to 62 %. This reduction does not only contribute to a reduced design time, but also the designer will benefit of the distribution system by reduced response times.

The estimation algorithm of FLODIS provides all information about the activities that is necessary to control the allocation algorithm. Until now, the allocation uses relative values (percentages) to indicate the resource requirements. The results show significant reduced response times of the activities. However, other allocation strategies using the available information may result in the same or more reduction of the overall execution time. An interesting approach is the evaluation of a more simplified estimation algorithm which classifies the activities. The allocation algorithm will avoid to select the same workstation for the same classes. On one hand, this algorithm needs less precise information about the activities, but on the other hand the savings of the distribution algorithm might be in the range of the originally proposed algorithm.

Acknowledgments

We would like to thank Prof. W. Rosenstiel and Prof. F. Wagner for the fruitful discussions and their encouragement of this research which enabled this publication.

References

[ARF90] W. Allen, D. Rosenthal, and K. Fiduk. Distributed Methodology Management for Design-in-the-Large. In *ICCAD*, pages 346–349, Santa Clara, 1990.

[Bus94] M. Bushnell et al. Distributed Computing, Automatic Design and Error Recovery in the ULYSSES II Framework. In *EDAC*, Paris, March 1994.

[CK93] T. Chiueh and R. Katz. A Distributed Transaction Facility for Design Task Management. In *EDAC*, Paris , France, 1993.

[DS66] N. Draper and H. Smith. *Applied Regression Analysis.* John Wiley & Sons, New York, 1966.

[ELZ86] D. Eager, E. Lazowska, and J. Zahorjan. Adaptive load sharing in homogeneous distributed systems. *IEEE Transactions on Software Engineering*, 12(5):662–675, May 1986.

[Kas92] Y. Kashai. Flow - A Concurrent Methodology Manager. In *EDAC*, Brussels , Belgium, March 1992.

[Kir94] S. Kirschke. Aufbau eines modularen, planenden Lastverteilungssystems für den Einsatz in Frameworks. Master's thesis, Universität Karlsruhe, January 1994.

[KS94] D. Kassabian and T. Soyata. *LB - The Load Balancer.* Dept. of Electrical Engineering, University of Rochester, April 1994.

[LJ92] D. Liebisch and A. Jain. JESSI-COMMON-Framework Design Management - The means to Configuration and Execution of the Design Process. In *EuroDAC*, Hamburg, 1992.

[NXG85] L. Ni, C. Xu, and T. Gendreau. A distributed drafting algorithm for load balancing. *IEEE transactions on software engineering*, 11(10):1153–1161, October 1985.

[TT85] A. Tantawi and D. Towsley. Optimal Static Load Balancing in Distributed Computer Systems. *ACM Journal*, 32(2):445–465, April 1985.

[Wei93] C. Weiler. Lastverteilung von CAD-Werkzeugen unter Einbeziehung von Framework-Informationen. Master's thesis, Forschungszentrum Informatik, November 1993.

PART SIX

Design Management

15

Design Flow Management: More than Convenient Tool Invocation *

Olav ten Bosch, Pieter van der Wolf and Alfred van der Hoeven

Delft University of Technology
Department of Electrical Engineering / DIMES
Mekelweg 4, 2628 CD Delft, The Netherlands

Abstract

The term design flow management is sometimes used for facilities that hardly offer more functionality than showing a graph of tool icons that describes the preferred order in which these tools should be executed. In this paper, we argue that a design flow management system (design flow system, for short) can and should be much more than that. If a design flow system supports the definition and visualization of data dependencies between tools, distinguishes between different design functions of a tool and guarantees the correctness of a design with respect to a configured design flow, it is more suited for use in a real-world design environment. In addition, a powerful design flow system should take the hierarchical decomposition and the different representations of a design into consideration and employ an intuitive mechanism for user-interaction. In this paper, we highlight these aspects of design flow management, based on the experiences gained by building design flow management in the NELSIS CAD Framework.

1 Introduction

In recent years, design flow management (sometimes called design methodology management) has attracted a lot of attention as one of the new technologies in the area of CAD frameworks. Although a variety of design flow systems with different capabilities have been presented (for an overview, see [1]), we feel that the benefit of using a design flow system is still underestimated. Design flow management can not only be used to invoke design tools in the right order, it can also be used to do *design tracking* (keeping track of the state of design and presenting it to designers in an attractive way), *constraint handling* (permit constraints on the design process to be defined and enforced), *guidance* (advise designers about further design steps) and *automation* (automatically execute parts of a design trajectory).

In this paper, we present our view on what functionality a powerful design flow system should have in order to be of use in a multi-tasking, distributed design environment, where multiple designers work with multiple design tools. We describe a number of aspects of design flow management to be taken

*This research was supported in part by the commission of the EU under ESPRIT contract 7364 (JESSI-Common-Frame).

into account when building a design flow system that has more functionality than just convenient tool invocation. For each aspect we describe its implications on the underlying flow model (i.e. the constructs to describe a design flow and the rules for design flow execution) and on the architecture of the design flow system (i.e. the design flow components and their integration into a CAD framework). We address the following aspects (there may be more, such as hierarchical flows):

- *data dependencies*:
  Temporal dependencies between design tools describe a design environment to a limited extent only. In addition to the order in which design tools should be invoked, it is also important to invoke them on data of the right type and on data that has the right state. The use of data dependencies in a design flow model increases the modeling power of the system, which permits a more accurate description of the design process.
- *fine-grain design flow management*:
  Instead of describing the design process in terms of tools, it is sometimes necessary to address functions of a tool. This yields a more detailed description of the design process. We describe how a design flow system can refine its description of the design process from a *coarse-grain* level, the level of tools, to a *fine-grain* level, the level of design functions of tools.
- *support for restrictive design flows*:
  Design flow systems that only use a design flow as a guideline for designers, are limited in the sense that they can not be used to configure and enforce design strategies. We claim that the enforcement of a predefined design flow can seriously improve the use of a design flow system without becoming too restrictive in experimental design stages.
- *complex design constraints*:
  For a seamless integration of design flow management into a design system, it is useful to offer design flow primitives that address other aspects of a design, like its hierarchical decomposition and its different representations. Such a design flow system can inform designers more precisely about the executability and successful completion of design activities. We describe a design flow model that supports the definition of design constraints that refer to the hierarchical multi-view structure of a design.
- *flow coloring*:
  An important part of the success of the design flow system is determined by its methods of user interaction. We present an approach that maximally exploits the configured design flow to present the state of the design process in an intuitive way to the end-users and that can be used in a concurrent design environment since it responds intelligently to changes in the design state.

To describe how these different aspects of design flow management contribute to a powerful design flow system, we take an incremental approach. We start from a 'minimal' design flow system and extend its functionality bit by bit to arrive at a design flow system that exploits the full potential of design flow management. Finally, we formalize the presented design flow model in a data schema.

2 A Minimal Design Flow System

The minimal design flow system from which we start is nothing more than a stand-alone framework tool that displays a graph of tool icons from which tools can be invoked. The graph expresses temporal dependencies between tools, which indicate the preferred order of execution. The flow model may use constructs like *and* and *or* to enable the definition of more complicated temporal relationships. An example of such a design flow is shown in figure 1. Tools are shown as rectangles and temporal dependencies are shown as arrows. This design flow shows a layout verification trajectory, with a layout editor, a design rule checker, a layout to circuit extractor, a stimuli editor, a simulator and a simulation result viewer. The **and** keyword indicates that the simulator can only be executed after an extraction has been performed *and* a stimuli description has been selected.

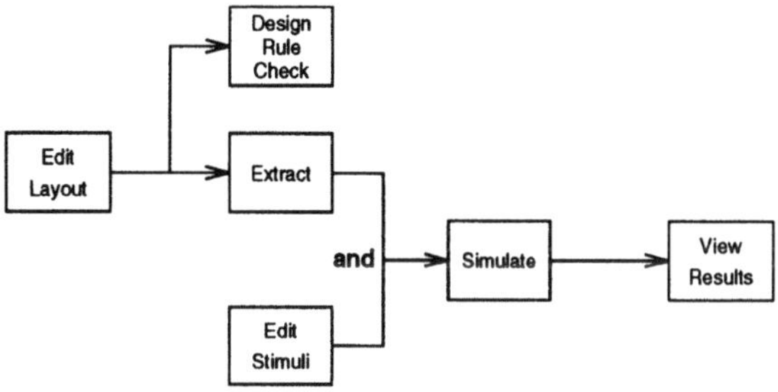

Figure 1: A design flow configured in a minimal design flow system.

3 Data Dependencies

A limitation of the minimal design flow system is that a design flow does not contain any explicit information on the data behavior of the design tools. It uses temporal dependencies, which express the executability of a tool in terms of the completion of tool runs. However, the executability of a tool may actually depend on a single piece of data produced within a tool run. Therefore, in order to describe the subtle dependencies among design tools with specific data access requirements, a design flow system should support the definition of *data dependencies*.

The addition of data dependencies to a design flow model does not mean that temporal dependencies should be discarded completely. They are inevitable to describe conditions that have no relation with data access at all. For instance, a requirement that an expensive system simulation may only be performed if some basic pre-checks have been performed, can only be described by a temporal dependency. A powerful design flow model should support data dependencies to guarantee the correctness of the design flow with respect to the data access requirements of design tools and temporal dependencies to allow the definition of conditions that do not directly originate from the data requirements of the design tools.

In order to support data dependencies tools can no longer be described as black boxes. Their behavior with respect to the data they access should be taken into consideration. The design flow model should be extended with a construct to denote data access or production. We use the term *design object* for a unit of design data for which the access and production is administered by the CAD framework and extend the flow model with a *port* construct (see also [2]) to describe access to design objects. A port is either an input port, describing read access to a design object, or an output port, describing the production of a design object. A data dependency can be described by connecting a number of output ports and a number of input ports to a *channel*. Temporal dependencies can be described in the same way as data dependencies, with the exception that no design object is read or produced via the ports of the channel. Figure 2 shows a design flow with data dependencies for the example design environment of figure 1. Ports are shown as small rectangles. An interesting observation is that the *and* construct in figure 1, which was needed to specify that 'simulate' can only be executed if 'extract' *and* 'edit stimuli' have been executed, is replaced by two distinct input ports for 'simulate', which determine its data access requirements. In fact, the *and* construct in figure 1 specifies the data access requirements of 'simulate', which are made explicit in figure 2.

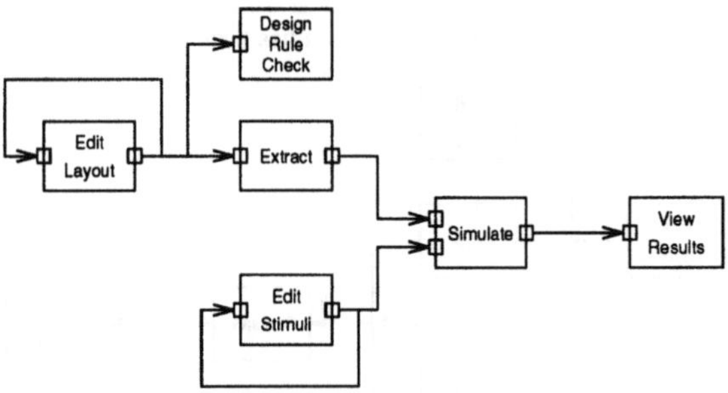

Figure 2: A design flow with data dependencies.

4 Fine-Grain Design Flow Management

For certain design applications, a description of the design environment in terms of tools and their data- and temporal dependencies, is still not detailed enough. For instance, in our example design environment, it may be that within one run of the layout editor multiple design objects can be edited. In this case, we would like 'edit layout' to refer to the editing of one individual design object in the design flow instead of the execution of the layout editor as a whole. Also, we may like to distinguish a flat layout to circuit extraction, which creates a flat circuit from a hierarchical layout, from a hierarchical layout to circuit extraction, which creates a hierarchical circuit from a hierarchical layout. For a hierarchical extraction, we would like to be able to describe its behavior on each level of the design hierarchy. This is visualized in figure 3.

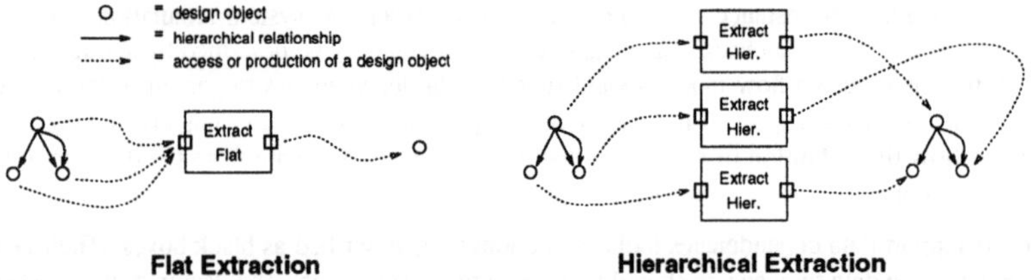

Figure 3: The operation of the extractor differs for a flat extraction and a hierarchical extraction.

In these examples, 'extract flat' and 'extract hier' are different *design functions* of the extractor. We say that a design flow system that allows tools to perform multiple, maybe different, design functions in one run and offers a mechanism to refer to these design functions from the design flow, performs *fine-grain design flow management.*

To support fine-grain design flow management, the design flow system should not only operate on the level of tool invocation and termination, it should be active during tool execution as well. In the flow model it must be possible to refer to the design functions of a tool. For this, we introduce the *activity* construct. (see also [2]). Several different activities may be defined for one design tool. A design tool may perform any of its activities any number of times during a single tool run. Figure 4 shows a design flow with activities for the example design environment of figure 1. A rectangle denotes an activity of a tool. Note that the input condition for the simulate activity, that *either* a flat *or* a hierarchical extraction has been performed *and* that a stimuli description must have been created, is modeled naturally using data dependencies.

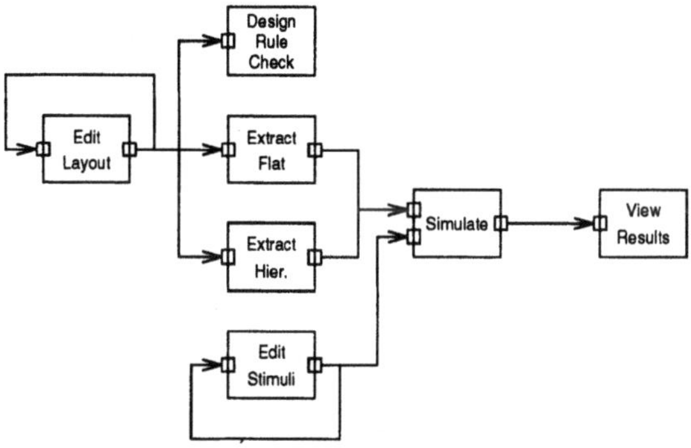

Figure 4: A fine-grain design flow with activities.

Besides a refinement of the design flow model, fine-grain design flow management has a strong impact on its place in the architecture of a CAD framework. To recognize individual activity runs within a tool run, the operation of a design tool needs to be monitored and compared with the configured activities for that tool. Figure 5 shows a general CAD framework architecture. The CAD framework (shaded components) consists of a set of framework services that guard the access to domain specific and domain neutral data, and a set of framework tools. Where a minimal design flow system could be implemented as a framework tool, a design flow system performing fine-grain design flow management must be integrated as one of the kernel framework services. In this way the design flow system can monitor all tools that access their design data via the CAD framework and perform design flow management on a fine-grain level. Support for interactive tools is then a natural fact. A detailed explanation of this material is given in [3] and [4].

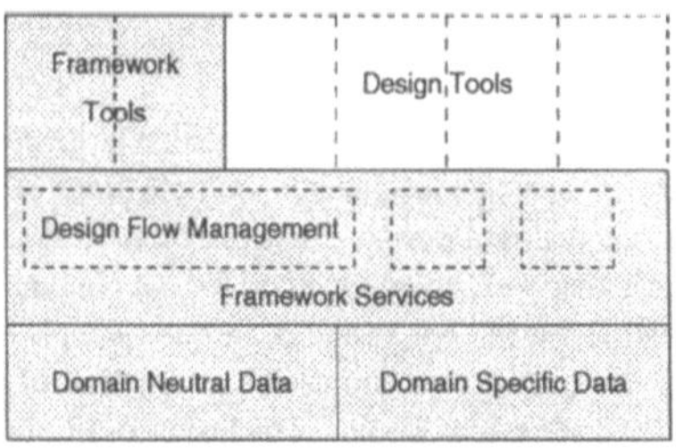

Figure 5: A framework architecture for fine-grain design flow management.

5 Support for Restrictive Design Flows

A common objection against design flow systems is that they would hamper designers in their experimental work. We stress that a powerful design flow system should allow the configuration of both *restrictive* as well as *unrestrictive* design flows. In this way a design flow system can be used for any type of design application and the designers' freedom of design is guaranteed. For instance, in the layout verification example a restrictive design philosophy may be required to demand that each layout must have been checked before it may be used in a flat extraction. A design flow for this situation is shown in figure 6. Note that, since no data is exchanged between the design rule check activity and the extractor activity, the channel between the design rule checker and the extractor specifies a temporal dependency. Also note that the design flow still offers alternative design trajectories as designers may still choose between a flat or a hierarchical extraction.

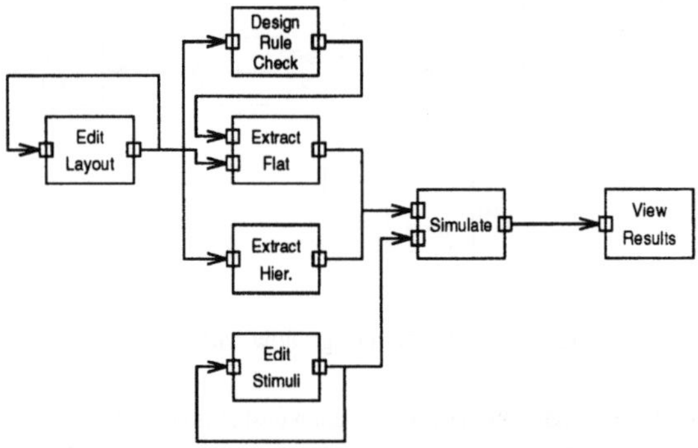

Figure 6: A design flow where each layout must have been checked before it may be used in a flat extraction.

If a design flow system supports restrictive design flows, this has implications on its implementation in the architecture of a CAD framework. In order to guarantee that tools started from outside the design flow user interface adhere to the configured design as well, the design flow system must be implemented in the kernel of the CAD framework. Therefore, the use of restrictive design flows is yet another reason to perform design flow management in the kernel of a CAD framework as described in the previous section (see also [3]).

6 Complex Design Constraints

In real-world design applications, it is quite common to organize large designs hierarchically, so that parts of the design can be described once and instantiated multiple times. This benefits the organization of the design data and facilitates concurrent design, since different designers may operate on different parts of the design hierarchy. Another well-known method for data organization is the use of different representations for a design, called views. For instance, in our example a design may have a layout view and a circuit view. The executability or successful completion of an activity may depend on the hierarchical multi-view organization of a design. For instance, in our layout simulation example we might want to require that a simulation may only be performed if the layout from which the input circuit has been derived and all its children layouts have been checked. We call such a design constraint that takes the hierarchical multi-view design organization into consideration a *complex design constraint.*

To support complex design constraints in a design flow system, its design flow model must offer constructs to refer to the hierarchical multi-view design organization. First of all, we need a way to refer to design objects that must satisfy a certain condition. For this, we introduce a special type of port, called a *condition port*. Second, we need a way to refer to relationships between design objects. For this, we introduce an *object relationship constraint* (ORC). As an example, a design flow with the complex design constraint that the simulator may only be executed on circuits if the layout from which the input circuit is derived and all the children layouts have been checked, is shown in figure 7. Condition ports are shown as small circles and ORCs as dashed edges connecting ports. ORC *a* is used to refer to the layout from which the circuit has been derived, ORC *b* to refer to all its children of this layout and ORC *c* to refer to the check results of these layouts. The upper condition port is connected to the output port of the design rule check activity, indicating that the check results must have been produced by the design rule checker. For a more detailed examination of complex design constraints, see [5].

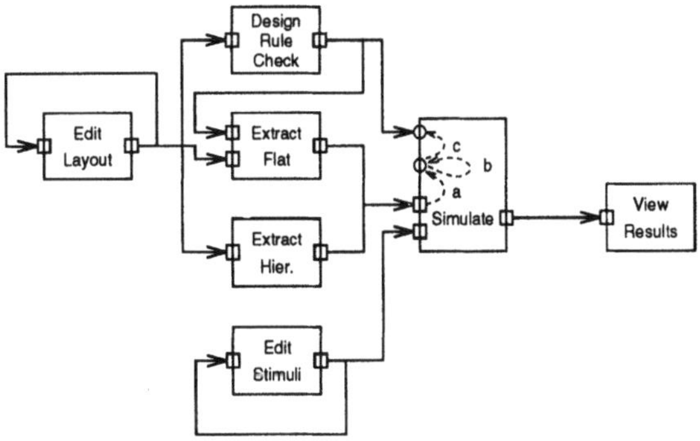

Figure 7: A design flow with a complex design constraint.

7 Flow Coloring

One of the fruits of a design flow system is that the state of the design process can be presented in relation to the configured design flow. Data and (activities of) tools can be visualized in one natural representation of the design environment. Crucial for the success of a design flow system is the method employed for interaction with the end-users. To be effective, it must display interesting parts of the design state and update the presented information when needed. A suitable mechanism for this is *flow coloring*. Flow coloring is the derivation of a *colored flow* that represents part of the design state. A colored flow is a design flow where channels and ports are 'colored' with a design object and activities are 'colored' with a state. This state denotes whether the activity is executable, has been executed or is not executable with respect to the design objects that color its ports. Flow coloring is the process of automatically deriving a colored flow starting from a *user-assigned* design object.

The colored flow to be derived represents the derivation context of the user-assigned design object. A number of *flow coloring operations*, like the assignment or removal of a design object to or from the colored flow, allow the designer to adjust a colored flow to his specific interests. An example colored flow is shown in figure 8. Colored channels are drawn thick and colored ports are drawn solid. Activities are thick if they have been executed, dashed if they are executable and dotted if they are not executable. In this example a layout has been generated with the layout editor, a circuit has been derived from the layout using a hierarchical extraction, and a stimuli description has been created. The simulate activity is not executable with respect to this data, since the layout from which the circuit has been derived has not been checked for design rule errors yet.

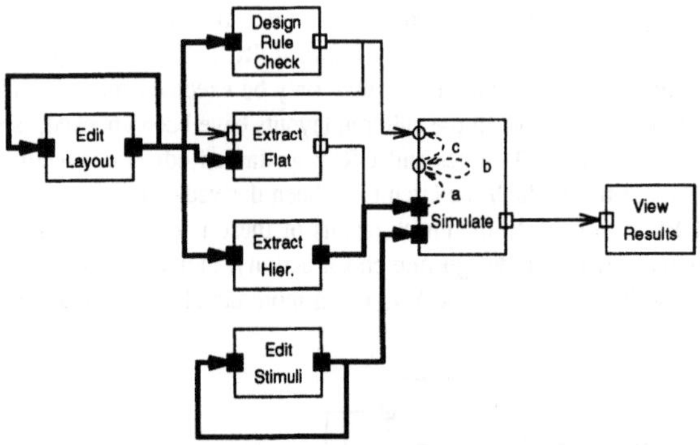

Figure 8: A colored flow, showing part of the design state.

A design flow user interface must be equipped with a facility to update the presented information upon changes in the design state. In a multi-tasking, multi-user design environment a designer may not want to be informed of design operations of its fellow-designers if they are not of his interest. In a user interface based on flow coloring, the information displayed in the colored flow is used to judge the designers interest in a change of the design state. If a change in the design state affects any of the design objects in the colored flow, the colored flow is recalculated and the designer is provided with a new colored flow. For a more detailed examination of flow coloring see [6].

8 Formalizing the Design Flow Model

Data modeling techniques permit a formal definition of the structural semantics of the information stored and maintained by a CAD framework [4]. In this section we use the OTO-D data modeling technique [7] to derive a *data schema* for the design flow model described in the previous sections. This data schema formally specifies the available design flow constructs and their relationships, and can be used to compare different design flow systems as has been done in [8]. The data schema, which is shown in figure 9, is a simplified[1] version of the NELSIS data schema. Boxes represent object types and lines connecting boxes represent attribute relationships.

The data schema specifies that an *activity* belongs to a *tool* and a *port* belongs to an *activity*. The *type* of a port describes whether it is an input or an output port and whether it is a condition port or not. Multiple ports can be *linked* to a *channel* to denote data and temporal dependencies between activities. An *ORC* is a directed binary relationship between a *Source-Port* and a *Destination-Port*. Multiple *activity runs* may be performed within a *tool run* of a *tool*. A *transaction* on a *design object* describes the access or production of a design object in an *activity run* via a *port* of the *activity* of that *activity run*. For each *transaction* holds that the *viewtype* of the *design object* involved in that *transaction* corresponds with the *viewtype* of the *port* via which that *design object* has been accessed or produced.

[1]For simplicity, we do not address the hierarchical organization of design flows in this data schema. The NELSIS data schema does support hierarchical design flows.

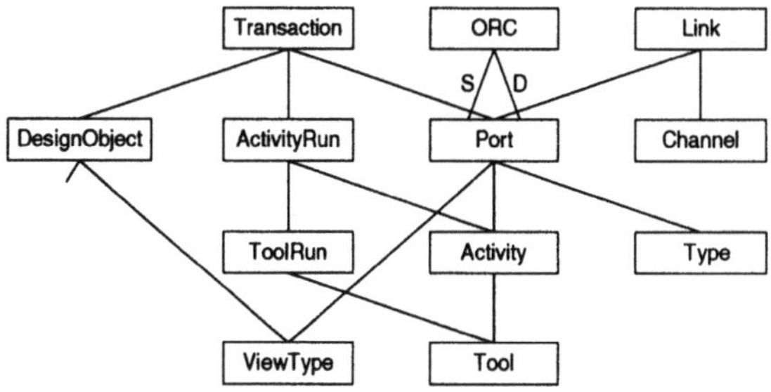

Figure 9: A data schema covering all aspects of design flow management addressed in this paper.

9 Related Work

Most design flow models described in literature do support data dependencies in some way. An exception is the Jessi-Common-Framework [9], where temporal dependencies with *and*, *or* and *xor* constructs are used to build design flows.

Examples of design flow systems that are capable to perform design flow management on a fine-grain level are VOV [10], which *traces* the data access and production of design tools and can therefore visualize changes in the design state while tools are still running and DECOR [11], which allows on-line monitoring of design actions.

Support for restrictive design flows can only be offered in those systems that have a notion of a predefined design flow, like the Jessi-Common-Framework [9], Hercules [12] and the system presented by van den Hamer et al. [13].

To our knowledge, no systems other than NELSIS address complex design constraints.

Flow coloring-alike paradigms have been used in other systems as well. For instance, in Monitor [14] a graph-like representation of the design flow was shown with traffic light colors, van den Hamer et al. [13] presented a user interface that shows data and tool runs in a *flowmap*, and in the Hercules system [12] a visualization of a *task graph* forms the basis of the user interface. Unique to our approach is that, starting from a user-assigned object, we derive a colored flow that represents those parts of the design state in which the designer is particularly interested and that we automatically adjust this colored flow upon changes in the design state.

10 Conclusion and Future Work

For various aspects of design flow management we described how a design flow system can be made suitable for use in a real-world design application. For this, a design flow system should use a design flow model that supports the definition of data dependencies as well as temporal dependencies, that is able to recognize different design functions of a tool performed during a single tool run, and that supports the definition of design constraints that depend on the hierarchical multi-view structure of a design. To perform design flow management on a fine-grain level and to support restrictive design flows, the design flow system should be integrated as one of the kernel framework services of the CAD framework. With respect to the interaction of the design flow system with the end-user, we propose a flow coloring paradigm, that uses the configured design flow to show parts of the design state in an attractive and comprehensible way to the end-user.

The presented design flow concepts have been implemented in the NELSIS CAD Framework to create a design flow system that has been used in a variety of design applications. Among the applications are a circuit design system, an analog system simulation environment, a design system for digital signal processor arrays, a synthesis system, a system for software testing, a simulation environment for medical applications and a generic system simulation environment. Future work will focus on how to deal with changing design environments.

References

[1] S. Kleinfeldt, M. Guiney, J. Miller, and M. Barnes. Design methodology management. *Proceedings of the IEEE*, 82(2):231–250, Feb 1994.

[2] K.O. ten Bosch, P. Bingley, and P. van der Wolf. Design flow management in the nelsis cad framework. In *Proc. 28th ACM/IEEE Design Automation Conference*, pages 711–716, San Francisco, June 1991.

[3] P. Bingley, K.O. ten Bosch, and P. van der Wolf. Incorporating design flow management in a framework based cad system. In *Proc. IEEE/ACM International Conference on CAD - 92*, pages 538–545, Santa Clara, Nov 1992.

[4] Pieter van der Wolf. *CAD Frameworks: Principles and Architecture*. Kluwer Academic Publishers, Boston/Dordrecht/London, 1994. ISBN 0-7923-9501-8.

[5] P. van der Wolf, K.O. ten Bosch, and A. van der Hoeven. An enhanced flow model for constraint handling in hierarchical multi-view design environments. In *Proc. IEEE/ACM International Conference on CAD - 94*, pages ?–?, San Jose, Nov 1994.

[6] K.O. ten Bosch, P. van der Wolf, and P. Bingley. A flow-based user interface for efficient execution of the design cycle. In *Proc. IEEE/ACM International Conference on CAD - 93*, pages 356–363, Santa Clara, Nov 1993.

[7] J.H. ter Bekke. *Semantic Data Modeling*. Prentice Hall, Englewood Cliffs, N.J., 1992. ISBN 0-13-806050-9.

[8] P. van den Hamer, K.O. ten Bosch, P. Bingley, M.A. Treffers, and P. van der Wolf. A comparison of two approaches to design flow management by data schema analysis. Technical report, Philips Research Laboratories & Delft University of Technology, June 1991. JCF Project Deliverable, SP1.

[9] D.C. Liebisch and A. Jain. Jessi-common-framework design management – the means to configuration and execution of the design process. In *Proc. EURO-DAC 92*, pages 552–557, Hamburg, Germany, Sept 1992.

[10] A. Casotto, A.R. Newton, and A. Sangiovanni-Vincentelli. Design management based on design traces. In *Proc. 27th ACM/IEEE Design Automation Conference*, pages 136–141, 1990.

[11] E. Kupitz and J. Tacken. Decor - tightly integrated design control and observation. In *Proc. IEEE/ACM International Conference on CAD - 92*, pages 532–537, Santa Clara, Nov 1992.

[12] P.R. Sutton, J.B. Brockman, and S.W. Director. Design management using dynamically defined flows. In *Proc. 30th ACM/IEEE Design Automation Conference*, pages 648–653, 1993.

[13] P. van den Hamer and M.A. Treffers. A data flow based architecture for cad frameworks. In *Proc. ICCAD - 90*, pages 482–485, 1990.

[14] A. Di Janni. A monitor for complex cad systems. In *Proc. 23rd ACM/IEEE Design Automation Conference*, pages 145–151, 1986.

16

Planning and Managing Multi-disciplinary and Concurrent Design Processes

Margarida F. Jacome
Electrical and Computer Engineering Dept.
University of Texas at Austin
Austin, TX 78712

Stephen W. Director
Electrical and Computer Engineering Dept.
Carnegie Mellon University
Pittsburgh, PA 15213

1 Introduction

It is time to move beyond the development of CAD tools that only aid designers in solving specific synthesis, analysis, and/or optimization design problems, but to develop tools that aid designers in planning and managing the increasingly complex design process itself. Specifically, we need to develop meta-tools capable of capturing design methodologies, and of intelligently making use of such methodologies while planning and managing the design process.

The more complex the design process is, the more useful can be a design process planning and management meta-tool capable of dynamically and coherently implementing an adequate design methodology throughout the entire design process. For instance, some design methodologies include mechanisms for controlling iterative design processes to ensure that complex, interdependent design constraints are satisfied. If, during the course of the design process, a constraint i s not satisfied, such design methodologies prescribe a backtracking strategy (i.e., indicate what design decisions should be re-accessed), based on the specific output produced by the analysis tools, and on the current coverage of the solution space (i.e., on the solutions already tried). In order to effectively explore methodologies embodying complex decision making, throughout the entire design process, meta-tools capable of creating an integrated representation of the complete design process are required. In this paper we will discuss the design process planning and management meta-tool Minerva, [1] [2], which is capable of capturing and dynamically implementing arbitrarily complex design methodologies.

This paper is organized as follows. In the next section we discuss hardware/software codesign, an area in which we feel that design planning and management is essential. Then, in Section 3, we point out the particular aspects of such a methodology that make it almost intractable, if only conventional CAD tools are available in the design environment. In Section 4 we present the formalism of design that constitutes the basis for Minerva. Then, in Section 5, we provide some details on Minerva's implementation, and illustrate some of its planning and managing services when supporting the codesign methodology described in Section 2. We finish with some conclusions in Section 6.

2 Example: Hardware-Software Codesign

In this section we review the h/s codesign methodology described in [3] to motivate the need for meta-tools for design process and planning.

Assume that we are given a behavioral description of a system we wish to realize. Such a description is typically a collection of processes written in a hardware description language such as Verilog, or a programming language such as C. The first step of the h/s codesign process is *behavioral partitioning, which* is aimed at determining which chunks of the original behavioral description, also called *tasks*, should be implemented in hardware and which chunks of behavioral description should be implemented in software. Observe that determining the granularity of the tasks themselves, i.e., the number and size of the hardware and software components, is also part of the behavioral partitioning problem. In general, partitioning is a highly *combinatorial* optimization problem. Moreover, even if a *satisfycing partition* (as opposed to an optimal one) is found that meets all requirements (e.g. speed and cost), these requirements can only be broadly *estimated* at the moment the partitioning itself takes place. This is so because at this point in the design process there is no concrete mapping of the resulting partition onto a specific design object. Thus behavioral partitioning is an *ill-defined* problem and it is very difficult, if not impossible, to develop *exact methods* for determining the best partitioning. Rather, *heuristic methods* have been developed to for determining an adequate partitioning in acceptable time. These methods work by dramatically pruning the search space but do not guarantee that a satisfycing partition, if one exists, will be found.

Given a suitable partitioning, the next step in h/s codesign is to group the resulting set of tasks into a new set of processes — those to be implemented in hardware and those to be implemented in software. This step is called *style selection.* As was the case with partitioning, style selection can only be achieved in a reasonable amount of time if heuristics are used to chose an adequate set of hardware and software processes.

After style selection, a *behavioral synthesis* step is used to realize the set of hardware components. This step, as with previous steps, is highly complex — recall that synthesizing an RTL implementation of an algorithm meeting speed, cost, and path requirements is an NP complete problem. Again, heuristic methods are used to overcome this difficulty.

To summarize then, to guarantee that a codesign process will be completed in an acceptable amount of time, heuristics must be employed. Unfortunately, heuristic methods cannot guarantee that satisfycing solutions, if they exist, will indeed be found. And certainly heuristics cannot guarantee that such a solution will be found on the first try.

Clearly the earlier a non-satisfycing solution is detected, the better. *Analysis* and/or *verification* tools can be used at various stages of the codesign process to assess the potential of a particular candidate solution to meet the design requirements. If the design is found to be non-satisfycing, the designer must *backtrack* (i.e., return to a previous design state), in order to try to adequately modify the design in order to bring it closer to a satisfycing one. Since the need for backtracking may occur repeatedly during a regular h/s codesign process, the h/s codesign process is iterative in nature.

When a non-satisfycing solution is detected, the designer must choose a *backtracking strategy* to determine which previous design state to return to, and then determine what should be modified. Properly choosing a backtracking strategy can be as critical to the efficiency and success of a h/s codesign process as is properly performing any of the "active" design steps (such as behavioral partitioning, or style selection). An example will illustrate this point.

Suppose that while performing behavioral synthesis for a particular hardware component the designer realizes that the resulting RTL description does not have the potential to meet one of the requirements. There are three backtracking alternatives: (1) re-synthesize the specific component, trying to impose adequate restrictions on the behavioral synthesis process; (2) modify the behavior for the particular component under design, by partially re-doing style selection; or (3) modify the particular tasks, by partially re-doing partitioning.

Observe that the nature of the requirements that are not satisfied, and the way these requirements relate to one another, determine which backtracking strategy to pursue. For instance, it is important to check if some trade-offs are possible among the candidate solution's current unsatisfied requirements. If such trade-offs are possible, this means that the current candidate solution may be quite close to a satisfycing one. Otherwise, it may be that the particular region of the solution space being explored does not hold any acceptable solutions. Observe that partially re-doing partitioning allows the designer to generate significantly different solutions, while re-synthesizing just the specific hardware component only allows the designer to move very locally in this space. Yet re-synthesizing is much less costly than partially re-doing style selection, which in its turn is much less costly than re-partitioning. Clearly, a more costly backtracking step should only be tried if the immediately less costly one proves to be incapable of generating a potentially satisfycing solution. Moreover, the solution space must be explored in a *controlled* and *coherent* fashion, such that, for instance, if a solution does not exist for a particular problem the designer is capable of identifying the situation in reasonable time.

3 Advanced Plan and Management Services

Let us highlight some key aspects of the codesign methodology described above. First of all, h/s codesign involves solving a significant number of complex design sub-problems, some of which may be addressed concurrently. Adequately planning a h/s codesign process thus involves choosing which methods and strategies to use for problem decomposition, during active problem solving, and then properly applying these for the solution of the codesign problem, while guaranteeing that consistency is preserved. As discussed below, maintaining consistency ultimately requires the ability to adequately characterize and handle sub-problem dependencies throughout the design process. Additionally, h/s codesign involves manipulation of quite different types of design objects, and is thus multidiscipinary in nature, thereby requiring a capability for handling, in an integrated form, the design of fundamentally distinct objects. Finally, h/s codesign is an iterative process. Adequately deciding on which backtracking strategy to pursue, given a specific failure, requires reasoning about the current overall state of the design process and, particularly, about the current coverage of the solutions space. It also requires, again, an adequate handling of sub-problem interactions, in the sense that such sub-problem interactions establish a causal chain between the various design representations and decisions created during the course of the design process. Such causal chains would significantly help in tracing back the design decisions that may have lead to a particular failure.

Summarizing, then, the effective application of h/s codesign methodologies requires a *unified* view of the design process, naturally integrating the design of objects of different nature, and the various kinds of problem decomposition mentioned in the previous section. Such a unified view is also needed to support decision making about backtracking strategies, and the actual implementation of such strategies. (Recall that the some of the strategies under consideration in h/s codesign processes may traverse several phases of the codesign process). Finally, guaranteeing consistency during active problem solving, as well as supporting reliable backtracking, requires the capture of sub-problem dependencies in this "unified" design process representation.

We have just characterized the basic infrastructure that is necessary for an effective implementation of the h/s codesign methodology described in Section 2. As will be shown in the next two sections, the Minerva design process planning and management meta-tool realizes such an infrastructure making it capable of fully incorporating and applying the h/s codesign methodology described in Section 2.

4 Minerva's Design Formalism

Minerva is based on a formalism of design that allows for a complete and general characterization of design disciplines, and for a unified, problem-based representation of design processes. Space precludes us from describing this formalism in detail. We will however introduce the principal concepts. Readers interested in a more detailed description should consult references [4] and [1].

Figure 1 illustrates the basic elements of Minerva's design process representation: design objects, design objectives, operators and consistency constraints. Objects under design, also called *design objects*, are represented and characterized by a collection of *properties* (or features). Such properties are partitioned by abstraction levels, creating one or more *facets* of the design object. Figure 1 shows a design object of the class "processing unit", with two facets: "algorithmic level," and "register-transfer level." The properties characterizing this design object (i.e., 'stored' in its two facets), are: "cost", "delay", "path", and "behavioral description." *Design objectives* define general classes of design problems. A general objective "synthesis" is shown in Figure 1.

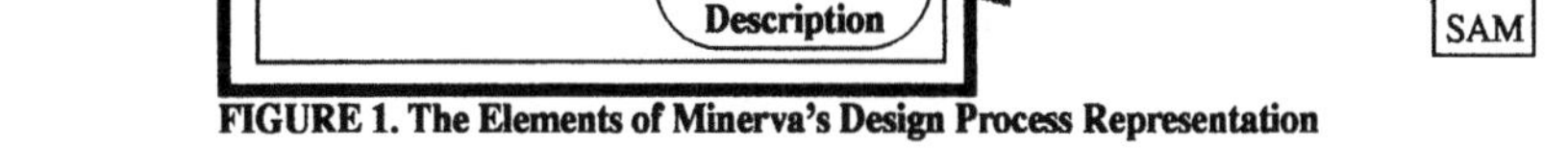

FIGURE 1. The Elements of Minerva's Design Process Representation

A design *problem* is defined in terms of a set of design objects (or, more specifically, a set of properties among those characterizing these design objects) and a design objective (viz. the problem to be solved within the context of the previous properties). Design problems are solved by *operators*, i.e., CAD tools.

So, in Figure 1 the CAD tool SAM [5] will be used to solve the behavioral synthesis problem defined by the design objective "synthesis" associated with a design object of the class "processing element" represented at "algorithmic level."

Problems can be directly solved if a suitable operator is available, and if the operator succeeds. Otherwise, an *impasse* has occurred. According to the nature of the impasse, it maybe resolved either by *problem decomposition* or by *backtracking*. Problem decompositions that take place during the design process generate a hierarchy of design problems where "parent" problems directly control their sub-problems. This problem hierarchy completely characterizes a particular design process. Figure 2 illustrates the fist level of problem decomposition for a h/s codesign process using the methodology described in Section 2. The control hierarchy is symbolically represented in Figure 2 by the arrows connecting "parent" objectives to their sub-objectives. (Observe how problem decomposition strategies can be easily captured and implemented in the high level, semantically rich problem-based representation of design processes.)

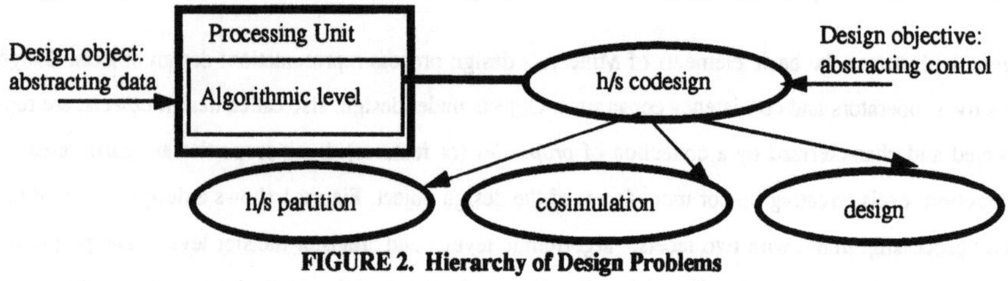

FIGURE 2. Hierarchy of Design Problems

Finally, *consistency constraints* express relations among properties, as symbolically illustrated in Figure 1. They constitute the basis for expressing sub-problem dependencies, as will be explained next. Observe that one of the forms by which problem decompositions make sub-problems treatable is by "artificially" assuming static values for certain properties (in the context of particular sub-problems), when indeed such values may yet not be static at all, or may even be impacted (to a certain extent) by the very decisions that will be made in the context of the particular sub-problem. This is the reason why, sometimes, sub-problem solutions may be inconsistent -- because the assumptions on the values of properties excluded from these sub-problem proved, later on, to be incorrect. Let us give an example. While implementing a particular chip component, the designer takes only into consideration the specific requirements stated for that particular component, such as the area reserved for it during layout planning. However, if the team of designers actually fails in generating all of the remaining chip components meeting their individual requirements, the design effort for the first component might end up being useless. We may say then that there is a "second order" dependency among some of such component requirements -- for instance, among the area of each chip component and the areas of the remaining components of the same chip. The properties "temperature" and "dissipated power" are another example of such a decomposition strategy -- their values tend also to be treated as independent (while addressing different design sub-problems) when in fact they clearly depend on each other.

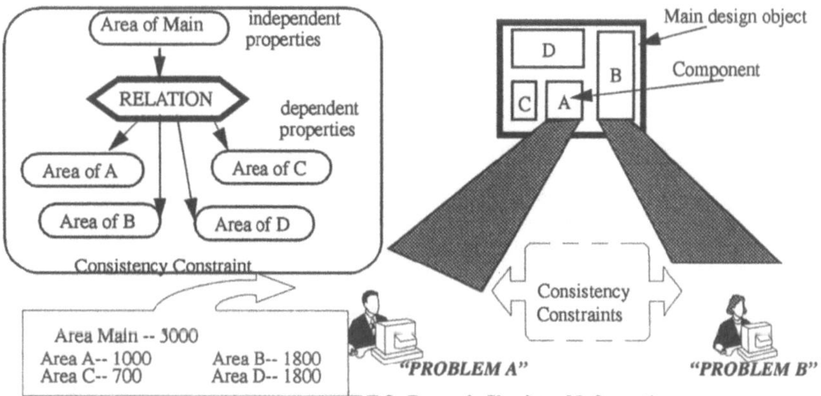

FIGURE 3. Dynamic Sharing of Information

Consistency constraints capture such "second order" relations among properties, and thus express sub-problem dependencies. Figure 3 shows a consistency constraint representing the relation between the areas of the components of a chip and the area of the chip itself. The semantics is as follows. Given a set of independent property values, and a relation, the set of dependent values must be chosen such that the relation stated in the constraint is verified. Otherwise, we say that the constraint was violated. Note that the later such inconsistencies are detected, the higher is the cost of backtracking (i.e., returning to a previous design state) in order to restore consistency, i.e., to adopt adequate assumptions on property values.

5 Minerva

In this section we discuss how Minerva's general capabilities adequately respond to the problem of planning and managing complex h/s codesign processes. We start by briefly discussing Minerva's general organization. As indicated in Figure 4,[1] Minerva has a set of basic components that serve all of its *sessions*.

1. Since space preclude us from presenting an exhaustive description of such an organization, in this paper we only address the fundamental aspects of it and remit interested readers to [2]

Such components include a *knowledge base*, that characterizes the design (sub)disciplines that are relevant to the particular design environment, and also an interface to an existing *executive meta-tool*, such as the one described in [6], that is responsible for executing CAD tools, storing design data, etc.

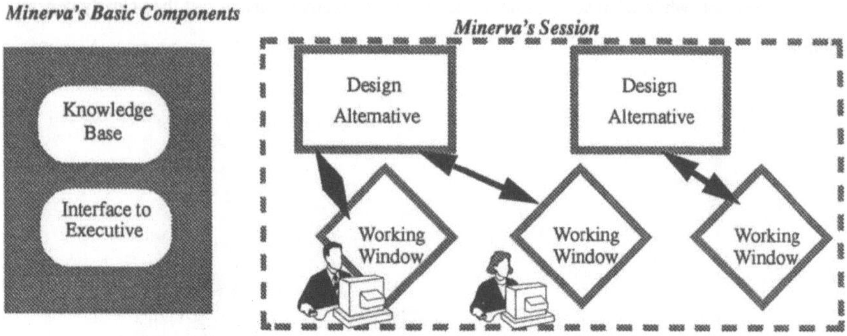

FIGURE 4. Minerva's General Organization

A session in Minerva, on the other hand, encompasses all of the *design alternatives* simultaneously under development for a given, central, design problem, as indicated in Figure4. Such a design problem can be, for instance, the design of a processing unit implementing a particular algorithm. Each design project ongoing in a particular design environment should thus have one Minerva session assigned to it.

Support of concurrent design requires guaranteeing that each design alternatives be *asynchronously* accessed by each of the designers assigned to the session's project. Figure 4 shows the general organization of a Minerva session -- in such an organization, one independent design alternative process exists for each solution being concurrently developed. Designers interact with these design alternatives through a dedicated agent called a *working window*. Figure 4 indicates that the design alternative depicted in the left is being concurrently accessed by two designers. "Working windows" and "design alternatives" cooperate (in a form that is transparent to the designer) in order to provide all of the synchronism features needed for supporting such concurrent design activity.

Figure 5 is a symbolic view of a hardware/software codesign process, as it would be shown to the designer through a working window. In this view, objectives already achieved are symbolically represented as dotted bubbles, while objectives not yet achieved are represented as solid bubbles. Objectives that can be concurrently addressed, at the current point in time, are shadowed. The symbolic view shown in Figure 5 corresponds to a snapshot of an advanced stage of a codesign process at which designers are already developing the set of hardware components that will implement the "hardware" tasks derived during the h/s partitioning (see Section 2).[2] Each of the shadowed "design" sub-objectives, together with their associated tasks

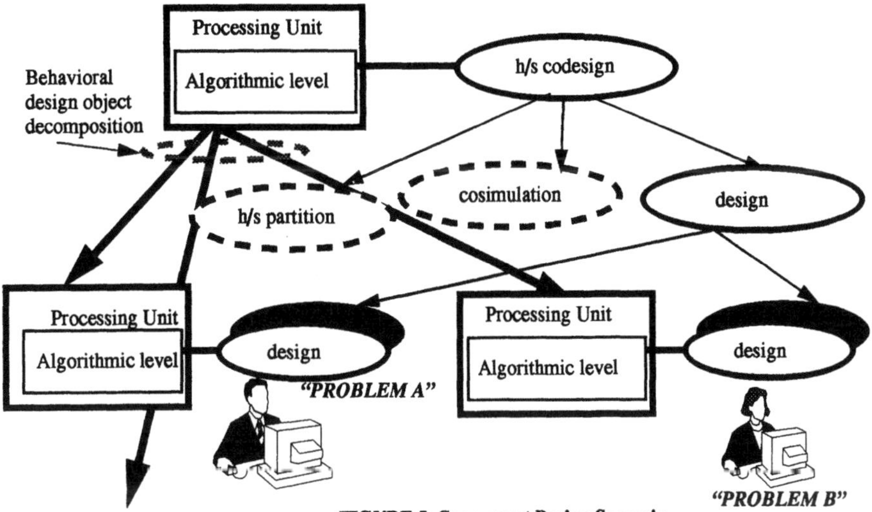

FIGURE 5. Concurrent Design Scenario

(stored within the processing units coupled to the sub-objectives), constitutes a design sub-problem that can be addressed concurrently at this stage of the codesign process -- in Figure 5 these sub-problems are labeled "Problem A" and "Problem B".

During active problem solving, when the designer selects a particular design problem to be addressed, the design information (i.e., set of properties) directly relevant for the accomplishment of the particular problem is delivered to the designer, through the working window. Additionally, Minerva's consistency constraints allow this same designer to dynamically monitor all of the properties that, not being directly relevant to the problem under consideration, may in certain circumstances have some impact on it, as discussed in Section 4. Figure 3illustrates this dynamic sharing of information that is provided in Minerva. Observe that the designer working on problem "A" is informed ("on-line") about the status of some of properties outside of the scope of this problem. Specifically, the designer is shown the amount of area being used by the designs worked on by other designers.

2. For simplicity purposes, a number of intermediate problem decompositions (mainly for the objectives h/s partition) are omitted from Figure 4.

During backtracking, consistency constraints are used by Minerva to show how dependent properties relate to their assumptions, i.e., independent properties. In particular, consistency constraints are used by Minerva to: (1) guide the selection of convenient backtracking points given specific failures; and (2) implement the backtracking steps themselves, i.e., propagating their effects throughout the design process. Let us give an example. Assume that in Figure 3the designer implementing "component A" concludes that the area originally planned for this component is impossible to achieve. The consistency constraint depicted in the figure, and all the others involving the particular design object failing to meet its requirements, will directly assist the designer in deciding: (1) what trade-offs can be tried -- for instance, what components could have their planned area reduced, in order to create space for Component A, or, (2) how the global problem can be redefined -- i.e. the requirements for the chip itself be changed. The concept illustrated in this example is generalizable to representing more complex constraints.

6 Conclusions

In this paper we demonstrated that progress in certain key design areas, such as those involving complex embedded systems, can only be significant if adequate design process planning and management meta-tools are available (together with the more conventional CAD tools). We then described Minerva, a meta-tool that is capable of capturing the fundaments of arbitrarily complex design methodologies, and of intelligently making use of such methodologies, while planning and managing the entire design process. We further discussed how Minerva naturally supports concurrent and multidisciplinary design, in the context of sequential or iterative design methodologies.

7 Bibliography

[1] M.F. Jacome, and S.W. Director, "Design Process Management for CAD Frameworks," in Proc. of *29th ACM/IEEE Design Automation Conference*. ACM Press, 1992.

[2] M.F. Jacome, "*Design Process Planning and Management for CAD Frameworks*," Ph.D. thesis, Carnegie Mellon University, Department of Electrical and Computer Engineering, September 1993.

[3] D.E. Thomas, J.K. Adams and H. Schmit, "A Model and Methodology for Hardware-Software Codesign," *IEEE Design & Test of Computers*, September 1993, pp. 6-15, September 1993.

[4] M.F. Jacome and S.W. Director, "A Formal Basis for Design Process Planning and Management", *Proc. of ICCAD Conference*, Santa Clara, CA. November 1994.

[5] R. Cloutier, and D. Thomas, "Synthesis of Pipelined Instruction Set Processors" In Proceedings of the ACM/IEEE 30th Design Automation Conference, ACM, 1993 (sam)

[6] P.R. Sutton, J.B. Brockman., and S.W. Director, "Design Management Using Dynamically Defined Flows," in Proc. of *30th ACM/IEEE Design Automation Conference*, ACM Press, 1993.

17

A User Model Supporting Communication in High-Level Design Systems

Anja Rockenberg, Susanne Heymann, Ursula Westerholz
Institute for System Design Technology
GMD-SET
Schloß Birlinghoven
D-53754 Sankt Augustin
e-mail: {rockenberg | heymann | westerholz}@gmd.de

1 Abstract

Today's high-level synthesis systems provide means to solve complex design problems but leave the user alone with an increasing number of difficult design decisions. We introduce a user model derived from requirements resulting from discussions with users and examination of existing systems. An implementation of this user model based on an existing high-level synthesis environment is sketched.

2 Introduction

Current frameworks and design systems administrate team work by providing intricate access and concurrency control mechanisms. In these models, the user is seen as an isolated person communicating with others exclusively through the use of common design objects. This view does prevent data collisions but it does not support human interaction in a working environment. Experience shows that good communication in a design team, the sharing of knowledge and results, increases work efficiency but on the other hand, people tend to work out a problem for themselves instead of asking a colleague who might already have found a solution. It follows that a design environment which takes user communication into account should provide means to share and distribute user experience within the system. This becomes increasingly important in high-level synthesis systems that require a lot of user interaction and expert knowledge in optimization steps.

We propose a user model introducing the concept of *virtual roles* to satisfy the need for communication between users in design frameworks. Our model is based on a small design team of up to eight people. Many state-of-the-art synthesis environments are tailored for single-user utilization, assuming a user that is well versed in all aspects of design. This is unrealistic – it is more likely that several people will pool their knowledge to achieve a design goal. On the other hand, due to the tight interaction between design steps, this group of people will not be very large.

This paper is structured as follows: We take a look at related work in the next section. This is followed by a definition of basic concepts for our work. Section 5 describes our sample system and the scenario from which we derive the requirements for our user model in section 6. A first implementation is presented in section 7. Our conclusions and an outlook to further work are presented in section 8.

3 Related Work

Cooperation of users in a design system is commonly seen as an access problem. First of all, users hold different levels of responsibility in their departments. These responsibilities vary according to the tasks a person is participating in and are reflected in access rights to task specific data. Secondly, while working together on a design, users are bound to work with tools that handle the same design data. Several approaches exist that address this kind of issue ([3], [1]). Our approach supports cooperation not just in respect of data management but on a more global scale. Conventional frameworks only consider conflicts as they appear; we introduce communication mechanisms that allow users to recognize impending conflicts and act to prevent them. If work in a small design team is to be partitioned effectively, this partitioning must be comprehensible to every individual designer. The design system must allow the designer to monitor progress of the complete task and structure his own work according to his needs. Software engineering environments like ADDD [6] support services like this.

Design management also deals with assistance in design flow issues and project management. The JESSI Common Framework provides the means to define projects and design flows ([7], [10]) that form a static work environment. Other tools merely record the design process to allow later replay or assistance in tool selection [2]. Predefined and prerecorded flows are a valuable assistance to designers because they provide a process template that precludes erroneous design steps. However, even in a static design flow there may be different paths to one design goal. Users – especially those not acquainted with the design system – will want to know which of the paths yields the best result. Other users' previous experience can help in this case if it is made available within the whole of a design task. We propose a new service that accomplishes this.

4 Tasks and Roles

A *task* is a set of design and management activities that lead to the accomplishment of a design goal. Within a task, activities may again be merged into *subtasks*. Since these may be divided to further structure the design process, a subtask hierarchy is finally achieved.

Users are assigned to subtasks. They may organize work within their subtask according to their own needs. Subtasks may be delegated from one user to another under certain circumstances.

Users working together on a task form a *team*. Every member of a team has a set of *roles* that define his rights and responsibilities within the task.
There are several possibilities for modelling roles. We realize roles as relationships between users and tasks.

5 Sample System Scenario

Our sample system consists of a high-level synthesis design environment, incorporating several synthesis tools and various optimization algorithms. The tasks of high-level synthesis systems are presented in detail in [3], [8], and [12]. We consider a small group of users that works in a tight team to design an embedded system.
The high-level synthesis system CASTLE (Codesign And Synthesis TooL Environment) [11] serves as a basis for this scenario. A very condensed form of the design flow is depicted in Figure 1.

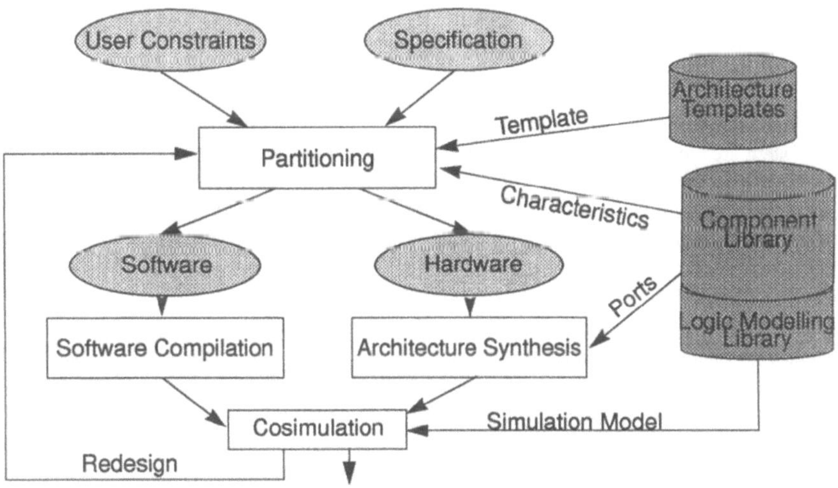

Figure 1. Prototype Scenario

We consider a team of two users working on the development of an MPEG video compression system. The functional description is available in C/C++. The conversion of this description to the CASTLE Synthesis Intermediate Representation format SIR is the first step of this task. This description is submitted to a profiling phase in the SIR/CASTLE Analysis Environment where information about operation count, instruction mix, most demanding functions, etc. is obtained. Based on these results, a design space exploration is performed and the principal system architecture is selected. In addition to this architecture template, a component library provides characteristics of processing components to fill the design with existing hardware structures.

This leads to the partitioning step where software and hardware parts are separated. During this process, the user should also consider information previously obtained by other users working on the same topic. Software is then directly compiled on the selected hardware architecture.

Additional application-specific hardware (e. g. ASICs, FPGAs) is synthesized and the joint result of these processes is subsequently verified by a cosimulation step using simulation models from the component library. The result of the cosimulation will normally lead to a re-partitioning of the system.

Processor synthesis and cosimulation are performed by commercial tools. Since this process is extremely time consuming, one wants to keep the probability of a redesign as small as possible. One means to achieve this in our small team is to have one designer perform compilation and synthesis steps while the other keeps trying to achieve better results in the analysis and partitioning step. If results that better match the design goal are found, they have to be communicated directly to the other team member, so that he may react accordingly.

Results are provided as a netlist of processing and storing components and their interface structure as well as the corresponding software components in C or Assembler.

A detailed design process using CASTLE is described in [14] and [15].

6 Requirements for Roles in a High-Level Synthesis Environment

In our sample environment, we consider a small team of up to eight users. Design steps in a synthesis system are closely interrelated: For example, any change in the partitioning decision immediately affects the software and hardware parts of the system to be built. In order to optimize the final design according to user constraints like performance or power consumption, several recursions of parts of or even the whole design process have to be performed. This process is highly interactive. Different optimization algorithms exist that may be chosen, and their results again depend on selecting the proper input parameters.

This close connection between design steps is reflected in the need for intensive cooperation between users because actions performed by one user will directly affect the results obtained by others working in the same task. Apart from the design-related work described above, there is the need for a coordinating agent. For our small team, we therefore derive the following roles:

- *Task Manager*: He supervises and coordinates activities within the design task. This includes the definition of subtasks and their delegation to designers as well as monitoring the progress of the task and its' results.
- *Designer*: Within his subtask, a designer must have as much flexibility as possible. He must be able to configure his environment, up to choosing his preferred tools or algorithms for a design step which offers alternatives in that respect. In addition, he may create his own subtask hierarchy to organise his work, delegate subtasks to other designers unless the task manager vetoes this action, and keep track of the progress of the whole task.

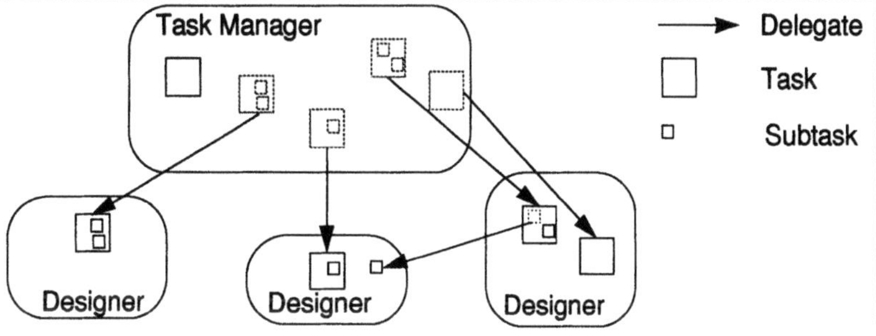

Figure 2. Task Structure

6.1 Virtual Roles

In addition to the roles fulfilled by real persons, we define two *virtual roles* that may be taken on by users as well as active system management objects. Virtual roles are initially delegated to the system but may be assigned to expert users who take over responsibility once they are chosen.

Virtual roles serve as key components in our approach and therefore require our special attention. They provide a simple and elegant means to model cooperation and communication as well as guidance in a design system like that described above where activities and their results are closely linked.

6.1.1 Moderator

The moderator is responsible for communication within a task. There is one moderator per task, responsible for recognizing communication needs and providing communication channels between users. A user must be able to tell the moderator that he does not want to be contacted directly at times. In that case, indirect mechanisms like mail may be used.

More specific, the communication aspects we envision with the moderator are organisational questions like task schedule, partitioning of a task, and the delegation of subtasks as well as conceptual questions like reactions to data collisions. In case of data collisions, new functionality for the solution of conflicts is provided. The moderator detects the need for communication in all of these cases and establishes the appropriate channels - either point-to-point or broadcast.

6.1.2 Adviser

The adviser provides an extended help service for users of the design system, ultimately combining help on framework services with application and design specific help. A hypertext-based, context-sensitive help server provides framework specific help. The integration of tool manual pages and help pages on design steps completes the system's part in this service. On request, default tool parameters and suggestions for design steps are provided, based on the evaluation of trace files. In addition, human users may register to the adviser as experts for information relevant to the task and system.

Contrary to common help servers, this mechanism provides the opportunity to reach information that is not written down formally but nevertheless may be important for accomplishing a design task. The adviser in this case points the user to the person who may be able to answer his question. This pointer may be directed to a specific help file maintained by the expert, giving hints to the user and protecting the expert from too many requests. If this help is not enough and direct communication is requested, the adviser signals the moderator to establish a communication channel to the expert who is contacted directly if he has not registered with the moderator as 'not available'.

7 Implementation of the User Model

Since the current CASTLE environment consists of stand-alone tools, our first implementation of the adviser and moderator roles is not based on an existing design framework. Instead, we use NCSA Mosaic [9] to display the adviser documents. The combination of Mosaic and HTML (HyperText Mark-up Language) easily allows to create a stand-alone adviser system. Forms are used for registering to the system as expert. Experts may insert their own documents into the adviser, realizing 'Frequently Asked Questions' pages that may be consulted before a human is actually contacted.

Further sources can be provided if the design system runs on internet-connected machines. In that case, access to the World Wide Web information services can be established.

The moderator role is realized as a suite of PERL [13] scripts that monitor adviser communication requests on the one hand and design file access on the other. Communication lines are established if the need is detected – either a button click in an adviser window or attempted access to a locked file.

8 Conclusions and Further Work

Acceptance of a system like that described above is difficult to measure! One major point of doubt was that users might refuse to register as experts to the adviser because they are afraid of constant interruptions of their own work as questions from others are flooding in. It turned out that in the small teams we have investigated here the psychological barrier to registering is quite low because people are well acquainted and need to work closely, anyway.

Novice users, e. g. students, needed less time to familiarize themselves with the CASTLE tool suite. Since the advisor/moderator duality is open and flexible to incorporate further domains, it can be extended to embrace advice on e. g. software engineering topics.

First results and feedback from users prompt us to extend the system. Next steps will be the integration of a WAIS database to better coordinate search for specific topics. In addition, we plan to integrate our service into an existing EDA framework.

It has been pointed out to us that the adviser service might be extended to encompass larger groups of users and a greater variety of topics. While this is technically possible we believe that there are limits to the amount of support that can be achieved. As the group of users accessing one adviser increases, so does the overhead for experts. It remains to be investigated how big the supported group of users may grow in order for the adviser to work sensibly.

9 References

[1] K. O. ten Bosch, P. Bingley, P. van der Wolf: "Design Flow Management in the NELSIS CAD Framework", *Proc. of the 28th ACM/IEEE Design Automation Conference*, 1991.

[2] A. Casotto, A. Sangiovanni-Vincentelli: "Automated design management using traces", *IEEE Transactions on CAD*, August 1993.

[3] D. Gajski, N. Dutt, A. Wu, S. Lin: "High-level Synthesis – Introduction to Chip and System Design", *Kluwer Academic*, 1992.

[4] A. van der Hoeven, O. ten Bosch, R. van Leuken, P. van der Wolf: "A Flexible Access Control Mechanism for CAD Frameworks", *Proc. European Design Automation Conference*, 1994.

[5] H.-U. Kobialka, C. Meyke: "Configurations are Versions, too", *4th International Workshop on Software Configuration Management*, 1993.

[6] H.-U. Kobialka, C. Meyke: "Views on an Object Oriented Software Engineering Environment", *Proc. of the 6th International Workshop on CASE*, 1993.

[7] D. C. Liebisch, A. Jain: "JESSI Common Framework Design Management – The Means to Configuration and Execution of the Design Process", *Proc. 1st European Design Automation Conference*, 1992.

[8] P. Marwedel, W. Rosenstiel: "Synthese von Register-Transfer-Strukturen aus Verhaltensbeschreibungen", *Informatikspektrum 15: 5-22*, 1992.

[9] The National Center for Supercomputing Applications: "About NCSA Mosaic for the X Window System ",
http://www.ncsa.uiuc.edu/SDG/Software/Mosaic/Docs/help-about.html, 1994.

[10] W. Schlegel (ed.): Design Management: JCF R4.0 Detailed Functional Specification – Draft Version, *JCF Document No. JCF/SNI/057-03/15-Jul-94*, 1994.

[11] M. Theißinger, P. Stravers, H. Veit: "Castle: An Interactive Environment for HW-SW Co-Design", *Proc. of the 3rd International Workshop on Hardware/Software Codesign*, 1994.

[12] R. A. Walker, R. Camposano (ed.): "A Survey of High-Level Synthesis Systems", *Kluwer*, 1991.

[13] L. Wall, R. L. Schwartz: "Programming perl – The Camel Book", *O'Reilly & Associates*, 1991.

[14] J. Wilberg, R. Camposano, W. Rosenstiel: "Design Flow for Hardware/Software Cosynthesis of a Video Compression System", *Proc. of the 3rd International Workshop on Hardware/Software Codesign*, 1994.

[15] J. Wilberg, R. Camposano, Ursula Westerholz, Uwe Steinhausen: "Design of an Embedded Video Compression System – A Quantitative Approach", *Proc. of the International Conference on Computer Design*, 1994.

PART SEVEN

Special Design Environments

18

A Framework for Electrical Test Engineering Automation

Gilson I. Wirth, Ricardo Wartchow, James Petterson and Sergio Bampi

Universidade Federal do Rio Grande do Sul, Informatics Institute, Caixa Postal 15064,
91501-970 - PORTO ALEGRE - R.S - BRAZIL
E-Mails: {wirth, wartchow, james, bampi}@inf.ufrgs.br

Abstract

This work deals with software environments for automatic test and characterization of electro-electronical devices, components and circuits. Special attention is paid to the features of testing and characterizing integrated devices and circuits.

The subject is treated in two different and complementary views:

i) The needs of the test and characterization engineer are addressed. The test engineer carries out physical experiments, which embody measurements and data acquisitions, and proceeds data processing, visualization and analysis.

ii) The needs of the software tools developer, who develops computer programs for the automation of the procedures that are usually carried out during test and characterization, are also addressed.

After the analysis of the subject under study, a software framework, called SGC ("Sistema de Gerenciamento e Controle"), was proposed and implemented. The SGC framework was implemented under MS-Windows using an object-oriented approach in a PC, a basic hardware in testing laboratories. The SGC framework aims to fulfill the needs inherent to the automatic test and characterization, when treated using the approaches mentioned above.

The SGC framework is an open system, supporting the easy integration of new software functions to the environment, as well as a practical system for test and characterization laboratory routines.

1 Introduction

To automate the testing and/or characterization of a device several tools are necessary. There are tools geared to specific tasks, namely: to control the test equipment and perform data acquisition, to display data, to analyze the acquired data, to perform parameter extraction, and so on. To increase engineering productivity, it is mandatory that these tools be integrated in the same test engineering environment. This environment can be called Measurement Information System (MIS) [1].

Several agents take part in a test/characterization process: people, measurement instruments, and computing tools [2], [3]. A test/characterization procedure is intended to be a description of a reciprocal action sequence of these subjects with the purpose of obtaining information on quantities that are inherent to a measurement object [1].

The user of a MIS must be able to: i) define the test procedure; ii) select the involved instruments; iii) supervise the execution of the test; iv) provide initialization parameters and (v) analyze the results through a user-friendly interface [3].

Other important features we have proposed and also implemented in SGC to increase productivity are: i) easiness to repeat or modify the procedures; ii) integrated documentation of the conditions in which the procedure was carried out; iii) generation of a procedures library; iv) support to data flow among different software tools; v) easy access to the generated/acquired data; vi) easiness to integrate/substitute tools and (vii) technology independence (the system should be well suited for all technologies used in the electro-electronics industry, like CMOS, Bipolar, ECL, InGaSP and others). These features should be readily integrated to the measurement software framework.

On the other hand, tool developers will have more productivity and produce better results if functions common to all tools and hardware interface management functions are provided by the MIS. Some of these functions are: i) data management; ii) hardware interface management (like RS232 or GPIB bus management); iii) tool intercommunication and (iv) user interface.

Several approaches to the problem have been proposed [1]-[10], while none of them satisfy all the above requirements.

The SGC framework was designed to fulfill all the above requirements, providing an efficient environment in which tools can be added by tool developers as well as they can be efficiently used by test engineers in routine tasks. In short, SGC provides for a test engineer the equivalent of an ECAD environment for a circuit designer.

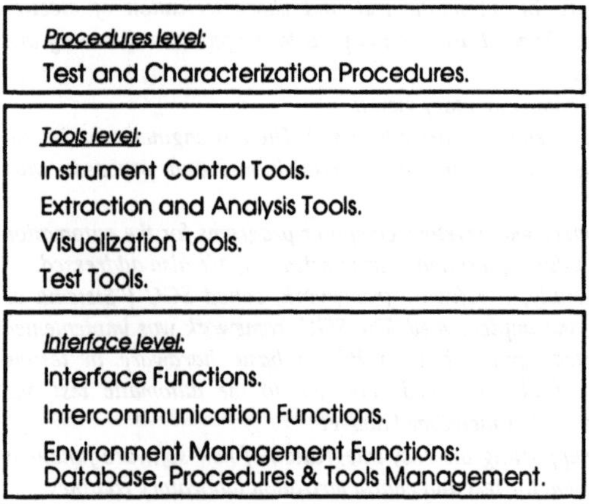

Fig. 1. Hierarchy of the SGC Framework

2 The SGC Modules

The main modules and functions of the SGC framework are organized in 3 levels of hierarchy: i) interface functions level; ii) tools level and iii) test/characterization procedures level; as shown in fig. 1. This hierarchical organization was done to provide the implementation of the components of a higher hierarchical level by repeatedly using the components of the lowest levels. Another goal is to decouple the components of the tools and procedures levels from the particularities inherent to the hardware were they are installed.

The object oriented paradigm was chosen to, jointly with the hierarchical organization, assure the modularity of the system. When the components of one level are changed, they keep the compatibility with the components of the other levels.

The software at the interface functions level is provided by the SGC as basic functions to communicate with the remote instruments, access the database and perform tool intercommunication, as well as manage the available software and hardware.

The components at the tools level are provided by tool developers whose task is geared to test, extract, analyze, display and acquire data.

At the test and characterization procedures level (higher level), the user can specify, modify and use specific testing procedures that repeatedly use the lower level tools.

The components of the three hierarchical levels are depicted in fig. 2. The components in gray boxes are functionally equivalent in all installations of SGC. The other ones may be different or even may not exist, because these modules are related to the particular hardware or test methodologies used. As one can see, all components are connected by means of the Intercommunication Module (ICM).

3 The Tools Level

At this level of the SGC structure, one deals with the tools needed to proceed the test/characterization procedures.

The tools can be divided in groups according to their functionality. The goal is to avoid the need to include the same facilities in different tools (e.g. to have visualization facilities in all tools). Every tool is designed to offer a particular facility, and using the set of interacting tools will address all user needs. The main groups of tools are: i) Instrument Control Tools; ii) Extraction/Characterization Tools (e.g. electrical parameter extractors); iii) Data Analysis Tools (e.g. visualization tools); and iv) Test Tools (e.g. test vectors generators).

When developing a tool, the software engineer should address only the facilities that are particular to the subject being modeled by the software tool. A tool to control a remote logic analyzer should only implement the facilities to control it and perform data acquisition with it. This tool should not address facilities to perform data display, test pattern generation or data analysis. These facilities will be addressed by specific tools. A tool used to implement a specific test vector generation algorithm, should model and implement just the specific test vector generation methodology it represents.

On the other hand, the particular facility being implemented should be made as generic as possible, not tied to a specific application (test/characterization methodology, although it models one). A tool aimed to perform display should be developed in a way that it can be used to visualize the data produced by a set of different tools.

The intercommunication facilities provided at the lower hierarchical level make it possible to transfer the data among the tools and use them in a complementary way.

This methodology intends to increase system flexibility, to save software development efforts, maximize the system modularity and make the user work easier (e.g., when data can only be visualized by means of the interface of the tool which has produced it, the user needs to learn how to use the visualization facilities of every tool).

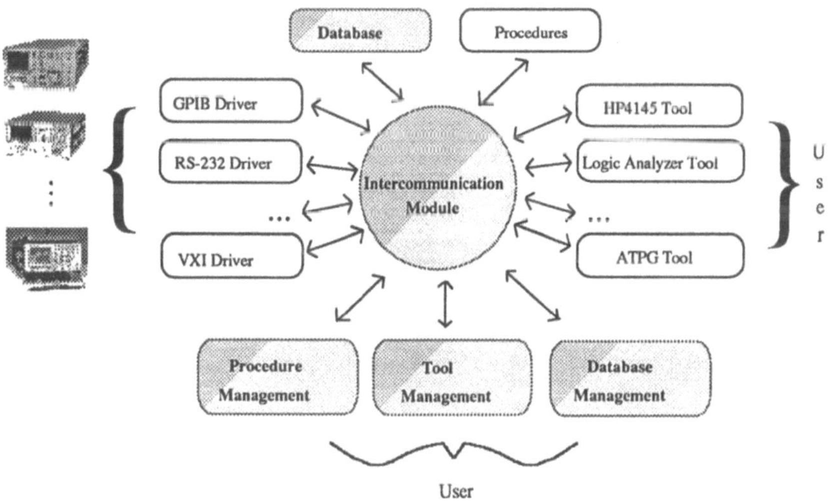

Fig. 2. SGC Framework Modules

4 The Procedure Management Module (PMM)

The PMM allows procedure retrieving, viewing, definition and execution.

The language used to implement the test procedure must: i) help in increasing the productivity of the test engineer; ii) be easy to learn; iii) impose no restrictions on the applications universe of the framework and (iv) be easily extensible.

Some MIS's use textual measurement programming languages (MPL) to allow the definition of a test procedure by the user [4], [5], [1]. But this language must be a minimal set of phrases covering all possible measurement problems in the given application area [1], thus imposing flexibility limitations to the applications universe of the MIS.

Other MIS's [3] implement the test procedure by means of the coding of software objects modeling the device under test and the chosen test procedure. To do this the user needs a good knowledge of the MIS implementation and good programming skills. In this approach, another problem is the documentation and repetition of the procedure, because many parameters have to be entered by the user in the initialization phase [3], and are not permanently saved, thus failing at points *i* and *ii* above.

The visual languages are the ones best suited to fulfill the requirements *i* and *ii* above [18].

In the SGC framework the user does not need to learn any textual programming language to proceed with the definition of the test procedures. All the steps of a procedure can be defined, viewed, edited and executed visually, through the visual primitives provided by the PMM and the visual user interfaces of the integrated tools. This implementation makes the user's work easier, guided by on-screen help.

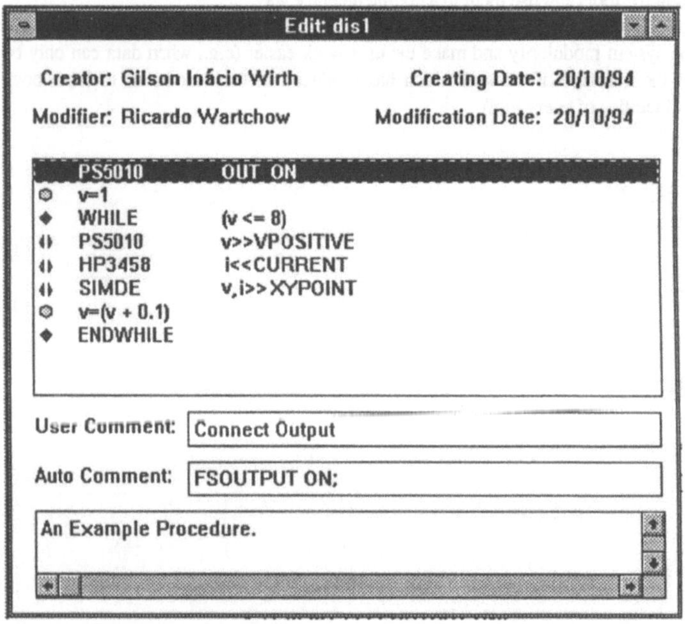

Fig. 3. Window for procedure visualization and edition

For the purpose of programming, a test procedure is sub-divided into steps. This sub-division can be better understood through an example. To characterize a ring oscillator with the supply voltage ranging from 1 to 8 V, in steps of 0.1 V, the following events may happen: i) the scale of a multimeter and a oscilloscope are set up (probably auto-scale will be chosen); ii) if the circuit is already characterized for VCC ranging from 1 to 8 V, the test is ended; iii) power is supplied to the circuit, by means of a programmable power supply, by setting up its output to the desired VCC value; iv) data like oscillation period, peak-to-peak voltage and rise time is acquired using the oscilloscope; v) the supplied current (ICC) is read using the multimeter; vi) if we want to visualize the ICCxVCC curve, every current value read by the multimeter must be plotted *versus* the value of the power supply output voltage; vii) if we want to visualize the dissipated energy per oscillation cycle *versus* VCC, the VCC value must be multiplied by ICC and by the oscillation period value; viii) the result of the multiplication in step *vii* is plotted against VCC; ix) the value of VCC is incremented by 0.1 V.

The steps of a procedure are classified as *setup, data flow, arithmetic, conditional* or *activity.*

A setup is a step in which the state of a tool changes, like steps *i* and *iii* in the above example.

An activity is a step in which a tool performs a task, like data acquisition (as in steps *iv* and *v* above) or data processing. Activities and setups are programmed through the visual user interfaces of the integrated tools.

In an arithmetic step arithmetic operations are performed over the data related to a procedure, like in steps *vii* and *ix* in the above example.

A data flow is a step in which data is transferred among tools or among tools and the SGC framework. This step allows the transfer of the data read by the multimeter and the voltage supplied by the power supply to the visualization tool that will plot the ICCxVCC curve in step *vi* above, for example.

Conditional steps are the ones in which decisions are made, like step *ii* of the above example. Data flow, arithmetic and conditional steps are programmed by means of the visual primitives offered in the PMM.

The description of part of the above example in the visual language implemented by SGC can be seen in fig. 3. In this example a power supply (PS5010), a multimeter (HP3458) and a visualization tool (SIMDE) are used.

Data flow steps in SGC are of the following types: *get from tool, get from tool and append, send to tool, send to database.*

Conditional steps are: *for loops, if then else statements* and the *while loop.* The dialog box used to program the conditional statements of these steps in a visual way is shown in fig. 4.

Arithmetic steps are *sums, subtractions, divisions, multiplications,* and so on.

This implementation of the procedure definition assures the best flexibility, not restricting the applications universe, because each object (tool) can be programmed on the basis of the knowledge it represents, and does not depend on the overall structure of the system or on a specific application (test procedure).

Integrating new tools to the framework can be seen as extending the lexical base of the visual language. After a tool is integrated, it is ready to be used to define new setups and perform new activities, adding new facilities to the framework.

5 Framework Management

Software modules at the Interface hierarchical level can be seen as framework management functions. These functions are used by all integrated tools, and include:

5.1 The Interface Functions

A tool under SGC never access a hardware interface like a GPIB, RS-232 or VXI interface directly. This access is implemented by means of *interface drivers* integrated to the SGC framework. The tool does not need to know to which interface hardware it is connected (like a GPIB interface used to access a remote equipment) or its configuration (like the address of a remote equipment on the bus or the end of string byte). All it needs to know are the data and the remote commands that must be exchanged with the equipment to perform its tasks. It sends a message to SGC requesting the sending/receiving of data or commands to/from the remote equipment. The SGC routes the message to the adequate driver, which performs the data/command transfer, managing eventual error conditions. The driver used and configurations like bus addresses are configured when the tool is integrated and can be changed at any time, as described in items 5.4 and 5.5.

In this approach a tool can concentrate on modeling the remote equipment it represents, and not the interface used to access it. Sometimes the same equipment is furnished in different configurations, using different hardware interfaces (like GPIB or RS-232), but the data and command strings exchanged are usually the same. Often GPIB interface boards used in computers communicate with application software in an incompatible way. In all these cases the software tools integrated to the SGC framework can be used without any changes. All that need to be done is to connect them to the adequate interface driver. New interface drivers can be integrated to the framework at any time, as described in item 5.5.

Already integrated in the SGC are drivers for the GPIB bus and the RS-232 interface.

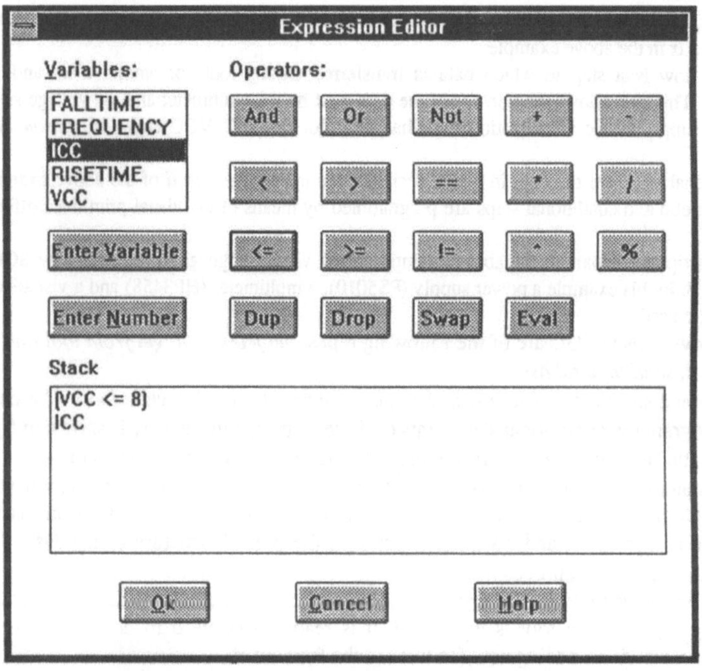

Fig. 4. Dialog box for the definition of conditional statements

5.2 The Intercommunication Module (ICM)

There are different needs of intercommunication in the scope of the SGC framework: i) the tools must inform SGC about the steps; ii) the SGC must ask the tools to perform steps when executing a procedure; iii) the tools must have access to the interface drivers; and (iv) there must be exchange of data, in real time or through the database; this data exchange must be possible even for foreign tools, which do not follow the SGC data models.

To fulfill these needs, three levels for intertool communication are provided by the ICM:

i) real time communication. This level implements the real time exchange of objects between tools, interface drivers and the SGC. It is implemented as a message passing protocol using the *MS-Windows* Dynamic Data Exchange (DDE) facilities [19].

ii) communication through the database. This communication is mainly used to exchange data among tools that are not active at the same time. Before a tool resumes execution, it may request the storing of the produced data in the database, making it available to another tools. The tools always send messages to the SGC requesting the storing or retrieving of data. They never access the database directly.

iii) use of translators to couple "foreign" tools with incompatible data type formats.

5.3 The Data Management Module (DMM)

The DMM provides an efficient and consistent way to store and retrieve the data acquired/produced, as well as the information needed to properly repeat and document a procedure. All the information related to the environment, like the tools used, its setup, sequence of the procedure steps and user comments are stored and related to the data acquired/produced in a consistent way. This avoids the spreading of lost data, whose acquisition/processing conditions, or sample device number under test, are not properly stored for later reference.

When the user wants to designate as input data to a tool data that is stored in the database, the tool exchanges real time messages with the SGC, informing the objects types it wants to receive. The DMM allows the user to proceed a browsing through the database, having as start point the procedures library or the integrated tools catalog, designating the desired data. The DMM assures that the designated data is in the format expected by the tool. The data transfer is performed automatically by the SGC DMM module, during the procedure execution, or immediately, when the tools are used in the stand alone mode.

The database is modeled according to fig. 5. For a question of brevity, it is not possible to discuss all the entities in detail. The main entities and its attributes are:

i) entity *Procedure*: defines the test characterization procedure and has attributes like a name, user comments for documentation, automatically generated comments, creation and modification dates, creator's and modifier's name. The entity Procedure can have N children, entities Setup, Activity, DataFlow, Arithmetic or Conditional.

ii) entities *Setup*, *Arithmetic* and *Conditional*: model the information needed to restore a tool setup or repeat an arithmetic or conditional step. They have attributes like a name (mainly for setups), identification of the related tool (only for setups), user comments and automatically generated comments for documentation. These entities have one an only one child, which stores the information that characterize the step.

iii) entities *Activity* and *DataFlow*: define the actions to be performed by a tool or data transfers to occur in a procedure step. They have attributes like a name, identification of the related tool(s), user comments and automatically generated comments for documentation. An entity which stores all information needed to repeat the step (*ActData* for Activities and *FlowData* for DataFlows) always exist.

One or more *PData* entities that are related may exist. They model the data produced by an Activity of a tool or transferred to the database in a DataFlow step. There are different data types defined in the SGC context (like vectors, matrices, SPICE parameter structures, test vectors, x,y points, etc.). Other data types may be defined to be used by the tools.

v) entity *Run*: models the execution of a procedure. It has attributes like an user comment, execution date and the name of the user who executed the procedure. The PData entities related to a Run are the ones produced by the execution of the Activity and DataFlow steps of a procedure.

vi) entity *Tool*: it models an integrated tool, and has relations to the steps that are defined when using it. It has attributes like the name to be displayed in the SGC tools menu, data types handled and the configuration needed for the communication through the related drivers (like GPIB addresses).

The database was implemented using C++ programming language and a relational approach.

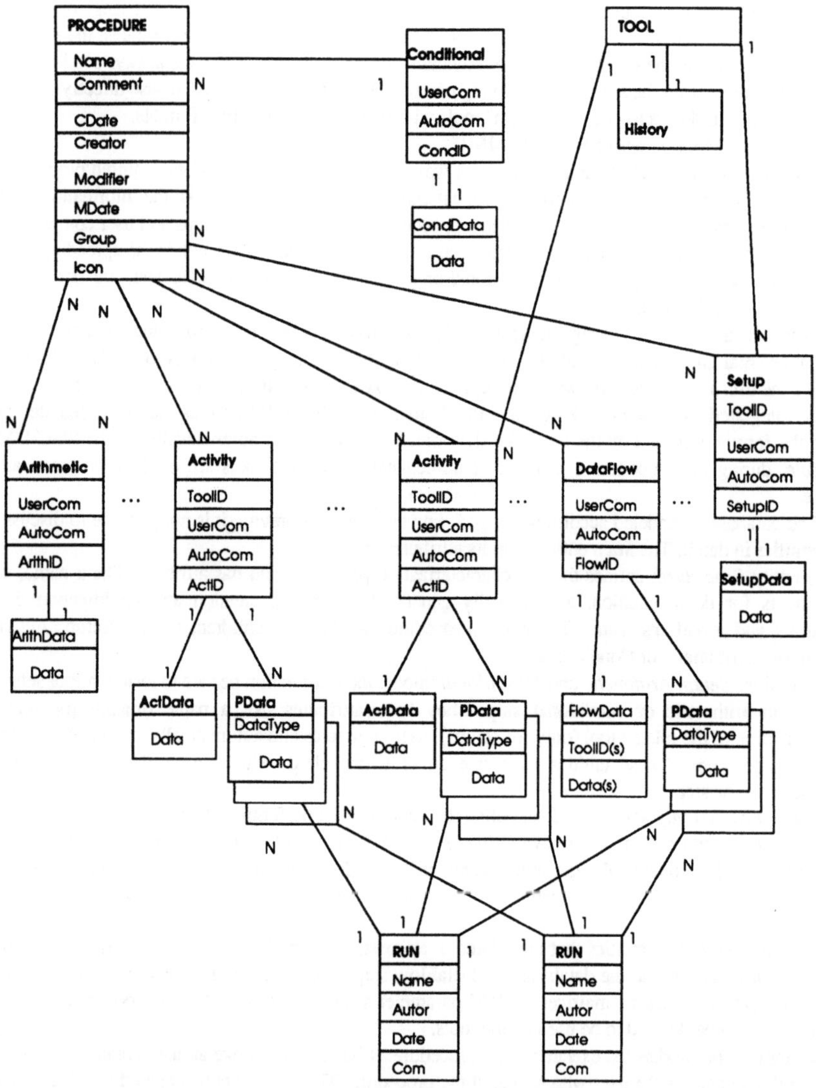

Fig. 5. The Data Model underlying the SGC framework

5.4 Tool Management

When the tools are not managed properly, even executing a tool can be a complicated task, if the user has to manually manage information like: i) tools name, version and physical location; ii) tools input and output data formats; iii) hardware needed to run the tool; iv) conditions to be satisfied before the tool runs; v) conditions generated after tool execution; vi) eventual use of interface drivers and their configuration (like bus addresses); vii) eventual need to execute data format translators; and so on.

SGC makes the user's work easier, managing all this information. To use a tool, all the user has to do is to select it with a mouse click at the SGC tools catalog.

Information like the tools and remote equipment history (vendor's name and address, contact person and phone, license expiration, maintenance information and so on) are also managed.

5.5 Tool and Driver Integration

SGC has specific facilities to allow the integration of tools and drivers, which significantly enhance the measurement productivity. Tools can be integrated by implementing the intercommunication protocol and supporting the data types defined in the SGC context.

To help tool and driver developers, skeletons are provided as C++ classes that implement, for tools: i) standard window structures; ii) callback and related functions to proceed the communication with all other framework components; iii) standard error handling; iv) basic functions needed for standard initialization and configuration saving/recovering; v) basic facilities for on-line help; and (vi) uniform dialog box examples. The driver skeletons implement: i) the basic structure for a *MS-Windows* driver; ii) basic functions needed for initialization and configuration saving/recovering; iii) callback functions for drivers and related functions to proceed the communication with the SGC framework; iv) handling for the standard messages; and (v) examples for the handling of particular messages and error conditions.

Instrument control tools as scopes, power supplies and function generators can be implemented starting from classes which implement the basic features. For example, a tool to control a specific scope model can be derived from a base scope class furnished by SGC. This base class implements the facilities that are common to most scopes. This feature is important because it allows the exchange and execution of procedures between laboratories that have different but functionally equivalent equipment (the base class stores and retrieves the information regarding steps related to them). This way measurement laboratories can exchange test/characterization methodologies.

SGC helps the user at connecting a new tool to the framework. Specific windows are used to connect it and perform the configuration of the parameters mentioned in item 5.4. Tool encapsulation (black box integration) is also available. This is done using a tool preprocessor and a tool post processor. The visual user interface of the tool preprocessor allows the definition of setups and activities to be performed and the data to be exchanged. When needed, the tool preprocessor runs data format translators. The tool post processor runs required data translators and communicates with SGC, transferring the produced data and reporting the status of the execution.

6 Future Work

EDA frameworks are widely used for the design of integrated circuit (IC) [11], and the IC produced needs to be tested and characterized. This test may be done at the end or during the production phase. The connection of the design environment with the test environment would greatly improve the productivity in the engineering environment. EDA tools provide, under the same framework, support to objects like chips, pins and test vectors. It is possible and desirable to exchange design data and their related test vectors to the test framework. These objects would be transferred to the proper test tool (e.g. a logic analyzer control tool). The result of the test could be visualized under a tool that could be common to both frameworks.

7 Conclusion

SGC provides a proper way for tool integration and module encapsulation, a consistent way for data storage, recovery and exchange, and has a friendly and well defined user interface. Thus, SGC is an appropriate environment for users and developers of measurement, test, extraction and data analysis tools for the automation of testing engineering routines, and although it was developed for use in our

Microelectronics Laboratory, the environment does not impose any limitation on the test applications universe.

We have been using SGC as a friendly environment for the test and characterization procedures done in our Microelectronics Laboratory. Also, test time was saved, because in microelectronics structures test/characterization, the procedures need to be done over a significant amount of samples, and repeating procedures under SGC is straightforward.

The tool development is easier now, because only the functions specific to a new measurement, analysis, test or extraction tool need to be coded. SGC showed its flexibility when we changed the GPIB board and handler, because no modification was needed in the Instrument Control tools, only the GPIB Interface Functions were modified, at the lowest hierarchical level of the SGC framework. The development of a MIS software framework for measurement management and tool integration proved to be a viable solution to the complexity of the tasks in the test/characterization laboratory.

8 References

[1] S. V. Muravyov and V. Savolainen, "Development Particularities for Programming Systems of Measurement Procedures". IEEE Trans. on Instrum. and Meas., Vol. 42, No. 5, pp. 906-912, Oct. 1993.

[2] A. Ferrero, "Software for Personal Instruments". IEEE Trans. on Instrum. and Meas., Vol. 39, No. 6, pp. 860-863, Dec. 1990.

[3] P. Daponte, L. Nigro and F. Tisato. "Object-Oriented Deseign of Measurement Systems". IEEE Trans. on Instrum. and Meas., Vol. 41, No. 6, pp. 874-880, Dec. 1992.

[4] C. Ringeard and M. Maloeuvre, "CALIBRAT; A Software for Control and Calibration of Electronic Measurement Devices". IEEE Trans. on Instrum. and Meas., Vol. 37, No. 6, pp. 497-500, Dec. 1988.

[5] "Automated Metrology", in Fluke and Philips Catalogue. Eindhoven, The Netherlands; Philips, 1990, pp. 220-238.

[6] Audio Precision System One, User Manual, Audio Precision, Inc., Beaverton, USA, 1989.

[7] E. G. Soenen et al., "A Framework for Design and Testing of Analog Integrated Circuits". IEEE Trans. on Instrum. and Meas., Vol. 39, No. 6, pp. 890-893, Dec. 1990.

[8] A. Caddemi, G. Martines, and M. Sannino. "Automatic Characterization and Modeling of Microwave Low-Noise HEMT's". IEEE Trans. on Instrum. and Meas., Vol. 41, No. 6, pp. 946-950, Dec. 1992.

[9] F. Russo and S. Broili. "A User-Friendly Environment for the Generation of High Portable Software in Computer-Based Instrumentation". IEEE Trans. on Instrum. and Meas., Vol. 39, No. 6, pp. 864-866, Dec. 1990.

[10] J. Doernberg. "Full-Speed Testing of A/D Converters". IEEE J. Solid-State Circuits, vol. SC-19, Dec. 1984.

[11] T. J. Barnes. Electronic CAD Frameworks. Kluwer Academic. Boston, 1992

[12] Yukata Ishikawa, "Communication Mechanism on Autonomous Objects". ACM Signlan Notices, vol. 27, nr. 10, pp. 303-314, october 1992.

[13] I. Videira and H. Sarmento. "Tool Integration Made Easier". Proceedings of the IFIP WG 10.2/WG10.5 Workshop on Tool Integration and Design Environ., 3., Amsterdam: North-Holland, 1992, pp. 207-224.

[14] E. T. Gutyet al. "Tool Communication in CFI Frameworks". Proceedings of the IFIP WG 10.2 Workshop on Tool Integration and Design Environments, 2., November 1990, Amsterdam: North-Holland, 1991, pp. 213-228.

[15] G. Kachel; E. Radeke and W. Heijenga. "Support of CAX-Applications by IDM, a Data Model of a Non-Standard Data Base System". Proceedings of the IFIP WG 10.2 Workshop on Tool Integration and Design Environments, 2., November 1990, Amsterdam: North-Holland, 1991, pp. 381-396.

[16] T. Kronmiller, "CAD Framework Integration: A Case Study". Proceedings of the IFIP WG 10.2 Workshop on Tool Integration and Design Environments 2., Amsterdam: North-Holland, 1991, pp. 1-8.

[17] B. Groening et al. "From Tool Encapsulation to Tool Integration". Proceedings of the IFIP WG 10.2 Workshop on Tool Integration and Design Environ., 2., Amsterdam: North-Holland, 1991, pp. 21-36.

[18] E. P. Glinert. "Visual Programming Environments: Paradims and Systems". Los Alamitos: IEEE Computer Society Press, California, 1990, 661 p.

[19] Microsoft Corporation. "Microsoft Windows Software Development Kit: Reference: Version 3.0". Redmond: Microsoft Corporation, 1990, 2 V.

19

A Compound Information Model for High-Level Synthesis

Peter Conradi
Center for Microelectronics
University of Kaiserslautern
Germany

Nikil Dutt
Information and Computer Science
University California at Irvine
USA

Abstract

High-level design and behavioral synthesis are the next natural steps in the evolution of CAD tools, dealing with the new generation of complex designs. Hence, a necessary future step in this progress of design automation is the standardization on more abstract levels of design. The basic problem is the lack of a good information model for high-level design. While the structure representation can be adapted from the CFI DR efforts, information models at the behavioral level are still not existent. This paper will outline the ingredients of a compact information model for high-level synthesis (HLS) from high-level languages such as VHDL.

Our contribution in this work is the presentation of an Information Model for HLS that captures the desired representations in a closed and minimized manner. The work is based on experiences with different HLS tools and algorithms at University California, Irvine.

1 Introduction

CAD research and development in the past two decades have resulted in fairly mature design tools and automation of several steps in the logic and layout design of electronic circuits. The CAD Framework Initiative (CFI, Inc.) has developed related framework and interoperability standards for these areas over the past 6 years. In January 1993, a first standard version for design representation was made available (DR 1.0). Within the CFI, different working groups and technical subcommittees are working on design representation issues for the logic and layout levels.

However, it is widely acknowledged that in order to cope with the explosion in design complexity resulting from advances in semiconductor manufacturing, the design process cannot be efficiently performed at the logic and layout levels.

Thus the interoperability demands for high-level synthesis have been growing over the last years, however without the necessary companion of suitable standardization work.

Since a number of years so called **Information Models (IM)** are used [1, 2] to explain relationships between different conceptional data entities in a formal and human readable manner. Thus standardization organizations like CFI, Inc. and EDIF agreed on the common IM language EXPRESS, developed by the STEP community.

Thus IM's for all EDA parts of design are essential for standardization and framework construction purposes. Using such an IM, procedural interfaces such as exchange formats or database schemes can be defined. In this paper we use the STEP-related format EXPRESS in its graphical form to present a common IM set for high level design and synthesis. The main problem in constructing and discussing an IM is to define a widely applied and efficient set of information used by the high-level design tasks.

2 Related work

Application topics for IM applications in CAD environments affect

- all tasks, which help the designer to **organize** his information space, version handling mechanisms, preparation for reuse, documentation etc.,
- all domains, where designers have to manipulate different **design descriptions** simultaneously (Gajski et al. [10], Rundensteiner [11], [12], Wu et al.[4]), and
- all design phases, where **different abstraction levels**, e.g., behavioral description, layout, technology information etc. are to be managed.

The user access in such a system could be done in a number of access modes that support specific needs on different views (black box information, floorplan information, technological conditions, RT-behavior, RT-structure, etc.).

Using IM techniques for design data abstraction has the following benefits: If the same IM is the base for concurrent software development, data adoption by each of the approaches will be easy. A database approach allows shared parallel development during component library design phase as well as shared usage of libraries. Object oriented treatment of component libraries and synthesis methods are state of the art. Links through different abstraction levels of a design enable reuse of design. The reason especially in high-level design is that use of components require knowledge about applicable operations. If the operations are not standard mathematical expressions, behavioral operation descriptions are necessary to meet the design goals.

Additionally, the technique of IM leads to more transparency of the design requirements in educational domains. The comparison and harmonization of information models of analog, digital, high-level and system level designs could offer more general insights into design processes.

For many years various types of tool integration have been suggested in the literature. The main problem of tool integration is the **definition of a unified data format**, that must be variable in size and volume in order to contain the required information.

There has been some related work in defining unified representations for HLS. For example, in the design trajectory from HL-design to layout, some authors propose unified representations [3, 4, 8]. However, these efforts were tuned towards synthesis and design support; they did not address the issue of defining a complete, minimal, and realistic usable IM for HLS. On the other hand, some authors are working on deriving information models for standard HDLs that support design capture and simulation [3, 4], but they do not address the issues of IM for the complete design trajectory including tools, algorithms, datastructures and libraries.

3 Compound Information Modeling

High-level design typically starts with the description of the design intent in a behavioral fashion (using an HDL such as VHDL), along with design constraints (such as area and power consumption) and overall design goals (e.g., maximize speed of execution). The designer has to specify a target library of components to allocate specific components. The high-level design synthesis process refines this behavior into an implementation composed of a structural netlist (datapath) of the allocated library components, sequenced by a Finite State Machine (FSM). A feasible output of the high-level design process is a complete FSM sequencing a

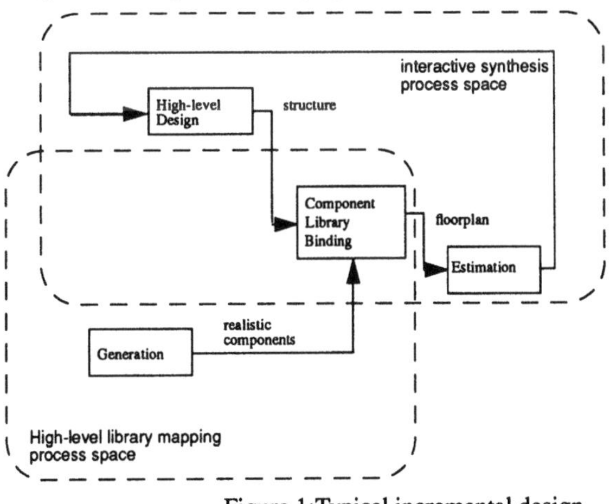

Figure 1:Typical incremental design process on system/high-level abstraction domains

datapath, that meets the design constraints, as well as the design goal. Thus, high level design is a repeated sequential process as shown in the SADT-diagram of Figure 1.

In this process the following representations have to be managed:

- Inputs:
  - Design behavior, expressed in a HDL (e.g., VHDL)
  - Design constraints
  - The goal functions for design
  - The resource or component library for refinement
- Internal Representation:
  - Control/data flow graph
  - Finite state machine or state table
  - Refined internal representation
- Outputs:
  - Structural netlist of library components (datapath)
  - Extended finite state machine (controller)
- Design Model: Finite State Machine with Datapath (FSMD)

In the complex multispace of behavior, structure and physical informations we are using a number of description methods. During this paper a **common view** of information will be constructed based on these diagrams.

We begin with the specification level, where no underlying realization model is existent. Specifying a high-level-design there are traditionally four kinds of different description methods. All these different representations could express different views of the same system object. While the **structure** net is a model of the units itself with their interconnection to other units, the **control flow** and **state transition diagram** show how the control flow is going between transitions; and the **data flow** diagram shows the activity of units and their flowing data.

High-level synthesis is based on specific design models, which are understood as the main frame for implementations. Following Gajski et al [9], the most suitable design model is the Finite State Machine with Datapath (FSMD).

In state-of-the-art high-level to layout synthesis, a set of all required design informations is to be stored for synthesis and reuse purposes. These different representations are linked after scheduling, binding, placement and routing tasks.

The basic design model used in HLS is the Finite-State-Machine with Datapath (FSMD) [9], where an FSM controls RT-level components in the datapath and sequences the design through various behavioral states. The FSMD model provides the link to the lower-level tasks of logic, layout and physical design. Based on the FSMD model, we found that the following kinds of information are collected together in order to find this complete set for an IM:

- Finite State Machine
- Scheduling information
- Data Flow Modeling
- Binding Information
- Structure representation
- Placement and routing information
- Layout configuration

4 Minimal Information Modeling

There is a trade off between describing data in a minimal form and conserving the semantic of the necessary description. Of course, it may be possible to describe all kernel data by configurations of nodes and links like in [4], but such a description has limited application scope. In our approach, the goal is to extract an IM of given programs, algorithms, and examples, which has the following advantages:

- defining a simple control flow model, which can be used for structured and unstructured control (see [7] for a definition) of parallel or sequential control, with different kinds of extensions,
- combining data flow and structure into one bipartite net, using the minimal additional effort to model the needed features.

The specific partial models will be described in the next sections.

4.1 Basic FSM

A HL design is typically specified in an imperative HDL (such as VHDL), and is refined into the FSMD model, where the control is embedded in a symbolic FSM. Hence it is necessary to contain at least this information.

We can represent a general FSM controller using the Mealy-FSM model. Figure 2 shows the information model:

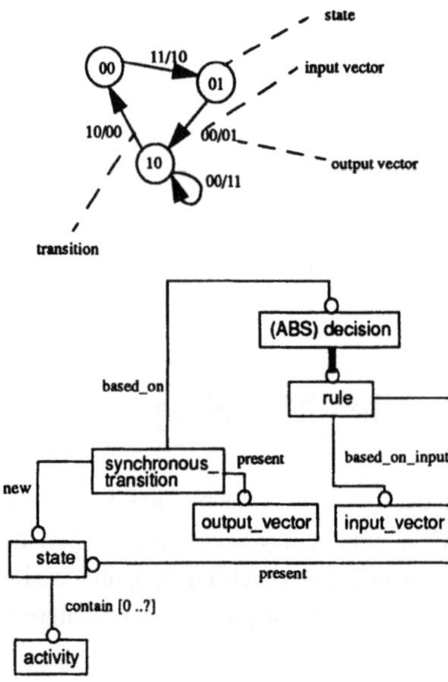

Figure 2: EXPRESS-G-diagram of
states and transitions

The information model consists of the entities *state*, *transition* and *rule*. Each *transition* is connected with a distinct *rule*, e.g. realized as a EXPRESS-function or a string, in the last case interpreted by database method. The execution of the *rule* offers the next-state- *decision*, based on the contents of two linked entities: *input_vector* and *state*. The execution of the database method results in a fixed link to one new *state* relation of the entity *transition*. An additional database method would be needed to copy the new *state* attribute from the decision-activated *synchronous_transition* to the *rule* in order to allow navigation through the FSM states.

Additional to the usual definition we define one or more entities of type *activity* as the task, executed during one state. With this option we have the capability to implement state-containing control and data flow parts.

Here we consider only flatted FSMs. If needed, composed FSMs could be modeled in a hierarchical way using the synthesis model of the FSMD, where each datapath component could hierarchically represent a lower-level FSMD.

4.2 Unified control

The behavior of the design can be expressed in the form of flows of control, along with associated data transformations. In a general context, these control transitions can be synchronous (i.e., transition on a clock), or asynchronous (i.e., transition on an event). Furthermore, control flow itself can be serial (sequence), parallel (fork), or conditional (select). The atoms in the control views are *activities*, interpreted as data flow blocks like shown in the next section. Hence, for an *activity* that contains all inner-state actions, we have to define:

- synchronous or asynchronous control, and
- conditional, sequential or parallel tasks inside of a state

An activity is controlled by a so-called (asynchronous) *transition*, a meta-control unit, which can be interpreted as a superset of a selection (*select*), a sequence (*sequence*), or a parallel control flow (*fork*), including the individual related join construct, like shown in Figure 3.

The *transition* bases on a *decision*, which could be a *rule*, modeled by an *activity* on a more detailed level.

In this way the capability to model structured control flow (CFG) and fork-join (FJG) constructs is given.

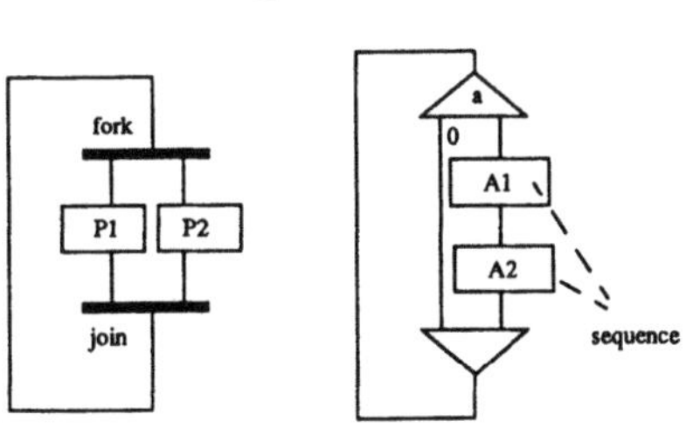

Figure 3: Different Control types

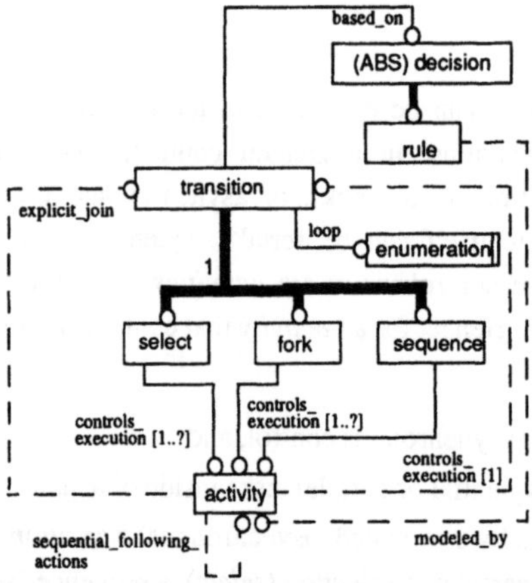

Figure 4: EXPRESS-G-diagram of unified control

Figure 2 and 4 have strong similarities. They can be combined quite easily by adding the *state*-construct to Figure 4. Also note that the *fork* construction in Figure 4 is equivalent to the Petri Net [13] form of control.

4.3 Structure, Data Flow

In a data flow view the same *activities* are containing scheduled Data Flow blocks. A Data Flow block contains operations such as READ and WRITE, algebraic operations and possibly decision nodes.

The complete Data Flow scenario could be understood as a interconnected network of activities by connecting affiliated write and read operations together.

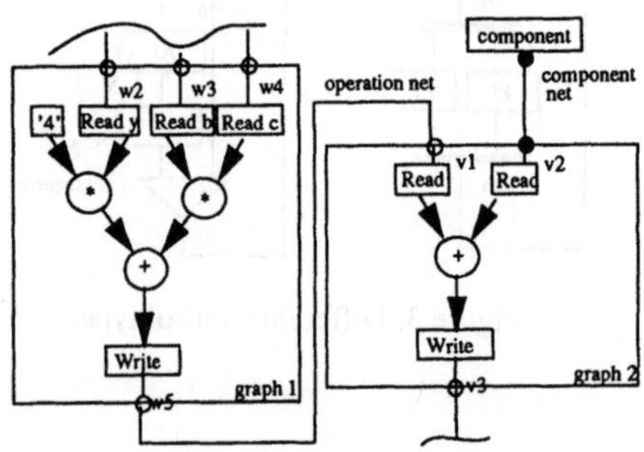

Figure 5: Dataflow graph and operation net

Exceptions are physical ports in Data Flow diagrams -- they are connected to real components; hence we have to combine the component-interconnections and operation-variables into one heterogeneous network.

For reasons of practical implementation, we propose to implement the Data Flow diagram in combination with the six-boxes connectivity model of the CFI, shown in Figure 6. In that way we understand operation nets as a partially coupled, but alternative representation of the design including the capability of defining relationships to component nets by bindings. Details of this model can be found in [6].

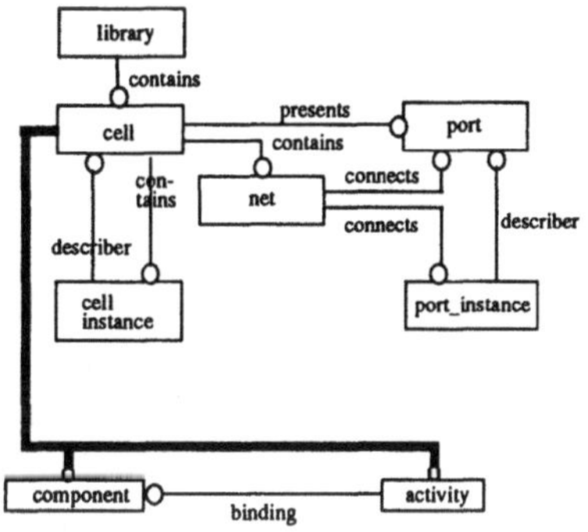

Figure 6: Extended CFI-DR model

4.4 Physical Layout

Floorplan and layout quantities will be described as physical location attributes, added to component and interconnection entities at detailed levels of granularity.

5 Summary

Our work is an initial effort towards the representation of a compound, integrated model of high-level design information. In particular, we presented a basic IM, consisting of

- integrated modeling of sequential, concurrent and conditional control flow
- combined data flow and structure modeling

This work was based on different experiments with design examples and synthesis tools, made at the University California, Irvine.

We believe this work complements the existing CFI work in the areas of DR, CIR, and VHDL IM. Future work will investigate applicability of this IM to a wider domain of HLS environments, as well as to its applicability in conjunction with existing IM efforts within a broader design context.

Literature

[1] Lau, R.Y.W.: "Proposal for an Information Model for EDIF", Technical Report # UMCS-91-6-2, Dept. Computer Science, University of Manchester, U.K., 1991.

[2] Marshall, M.A.J., Kahn, H. : "A Structural Information Model of VHDL", in Mermet, J., (ed.), "VHDL for simulation, Synthesis and Formal Proofs of Hardware", Kluwer Academic Publisher, 1992.

[3] Knapp, D. W., Parker, A. C. : "A unified Representation for Design Information", Computer Hardware Description Languages and their Applications, IFIP, Elsevier, 1985.

[4] Wu, A.C.-H., Hadley, T.S., Gajski, D.D.: "An efficient Multi-view Design Model for Real Time Interactive Synthesis", Proc. ICCAD, 1992.

[5] CFI: "Design Representation, Electrical Connectivity Information Model and Programming Interface", CAD Framework Initiative Inc., Version 0.9.4-071092, 1992.

[6] Conradi. P.: "Information Analysis in High-level Synthesis", Technical Report # 94-39, Dept. Computer Science, University California, Irvine, 1994.

[7] Ang, R. P., Dutt, N.: "Equivalent Design Representations and Transformations for interactive scheduling", Proceedings of ICCAD, pp. 332 - 335, 1992.

[8] Chaiyakul, V., Gajski, D.D.: "Assignment Decision Diagram for High Level Synthesis", Technical Report # 92-103, Information and Computer Science, University California, Irvine, 1992.

[9] Gajski, D.D., Dutt N.D., Wu, A.C-H, and Lin, S.: "High-Level Synthesis: Introduction to Chip and System Design", Kluwer Academic Publishers, 1992.

[10] Gajski, D.D., Vahid, F., Narajan, S., Gong, J.: "Specification and Design of Embedded Systems", Kluwer, 1994.

[11] Rundensteiner, E., A., Gajski, D.D. :"A Design Representation model for High-Level-Synthesis", Technical Report, University California at Irvine, 1990.

[12] Rundensteiner, E.: "Object-Oriented Views: A novel approach for Tool Integration in Design Environments", Technical Report # 92-83, Information and Computer Science, University of California, Irvine, 1992.

[13] Peterson, J.: "Petri Net Theory and Modeling of Systems", Prentice Hall, 1981.

20

An object–oriented CAE Framework for asynchronous Circuit Synthesis

U. Baake and S. A. Huss

Integrated Circuits and Systems Lab.
Department of Computer Science
Technical University of Darmstadt
64283 Darmstadt, Germany
{baake, huss}@vlsi.informatik.th–darmstadt.de

Abstract

The paper presents an interactive OSF/Motif–based design framework for high–level specification, analysis and synthesis of asynchronous control circuits. As novel contributions we propose an object–oriented graphical design environment including interfaces to external state of the art asynchronous circuit synthesis tools. The proposed design environment represents a powerful CAE design system including features like geometric representation of abstract graph–based input specifications, menu–based command entry, user–defined configuration of input devices and a comprehensive set of visualisation and manipulation commands. Thus, by using this system, a circuit designer is able to specify and visualise asynchronous circuit behaviour on a high abstraction level and synthesize the corresponding logic using the built–in algorithms as well as other available non–graphical synthesis tools.

1 Introduction

In the domain of interface circuitry, where the use of a global clock is rarely applicable, design methods for asynchronous circuits have recently gained an increasing interest in the logic synthesis community [LL91][LS93][Men90][MM94]. By using design methodologies for locally synchronous and globally asynchronous operation of data processing [Bor89] problems related to limited throughput, clock skew, and clock distribution may be avoided. Even in this restricted domain, asynchronous design has been considered as time–consuming and difficult because of the lack of both, a good formal specification and of appropriate synthesis and verification tools. Formal specification of asynchronous circuits is not that easy because behaviour such as concurrency, sequencing, conflict, timing constraints and data–dependency is difficult to specify in a way that is both, natural for designers and at the same time appropriate for formal analysis or verification. In contrast to synchronous design, where powerful CAE environments, such as those provided by, e.g., Synopsys or Viewlogic, support the specification and synthesis process, available asynchronous circuit synthesis tools, such as e.g. SIS [BLM+92] from UCB or ASSASSIN [YVL93] from IMEC, use an alphanumeric design entry only.

In this paper we present an object-oriented, interactive design environment that supports automated asynchronous circuit design starting from a behavioural graph–based problem specification called signal transition graphs STGs [Chu87]. This formal specification form is exploited in other well–known synthesis tools, e.g. [BLM+92], too. Thus, the implementation of well defined interface layers allows specification-import and data-export between the proposed framework and e.g. SIS or ASSASSIN. A powerful component for visualizing and beautifying graph–based specifications allows for a comfortable processing of imported non–graphical STG specifications. Using the OSF/Motif–based graphical front–end of the framework for flexible design capture and manipulation, a designer can use proprietary algorithms for analysis or synthesis as well as other well–known synthesis methods by exporting the specification according to their requirements.

In addition to the object–oriented representation of STGs [BH93], we introduce algorithms for constructing geometric representations from textual specifications. This is essential for importing and visualizing graph–based textual specifications within the graphical front–end of the framework. The automatic generation of drawings of graphs has important applications in key computer technologies such as software engineering, database design, and visual interfaces.

The paper is organized as follows. In Section 2, the specification form used for describing interface behaviour is briefly introduced in a formal way. In Section 3, we give an overview for the main components of our system, called Graphical Asynchronous Circuits Environment. In Section 4 a more detailed description of the GRACE front–end is outlined, while Section 5 describes the analysis and synthesis parts, respectively. In Section 6 the import and export interfaces of GRACE are described. An example of a complete design flow, processed by this CAE framework, concludes the paper.

2 Design specification

In this section we give a brief introduction into signal transition graphs and show that this specification is very well–suited to describe asynchronous circuit behaviour. STGs, introduced independently in [Chu87] and [RY85], represent a graph–based methodology for the specification of speed-independent asynchronous digital controllers. An STG provides an appropriate representation of concurrency. In addition its natural timing–diagram–like specification makes it easy to describe interface communications with signals as outlined in Figure 1.

An STG $\Sigma = \langle T, P, F, m_0 \rangle$ is an interpreted Free Choice Net (FC Net, i.e. subset of a Petri Net), which is expressive enough to specify concurrency or choice, but yet simple enough for analysis. The set of transitions T is interpreted as the set of signal transitions $S \times \{+, -\}$. S denotes a set of signals, where for every signal $s \in S$ a rising transition s^+ and a falling transition s^- is associated, respectively. s^* denotes any transition of signal s. P is a set of places that can be used to specify conflict or choice and $F \subseteq (T \times P) \cup (P \times T)$ represents the flow relation between transitions and places. The marking is an integer labeling of the places $m : P \rightarrow \mathbb{N}$, denoting the number of tokens in a place p. m_0 represents the initial marking of Σ. The *preset* of a place or a transition $x \in P \cup T$ is given by $\cdot x = \{y | (y, x) \in F\}$ and the *postset* of x is denoted $x\cdot = \{y | (x, y) \in F\}$, respectively.

A transition t is said to be enabled, if $\forall p \in \cdot t : m(p) \geq 1$. This is denoted as $m[t\rangle$. The set of all enabled transitions is then $\mathcal{F} = \{t \in T \mid m[t\rangle\}$. An enabled transition may fire according to the Petri Net firing rule.

A marked graph (MG) is a Petri Net, where each place p has exactly one predecessor transition, denoted as *source*(p), and one successor transition, called *target*(p). It is usually represented by a directed graph G, where the vertices correspond to transitions and the arcs to places.

An STG/MG is an STG that is derived from an MG.

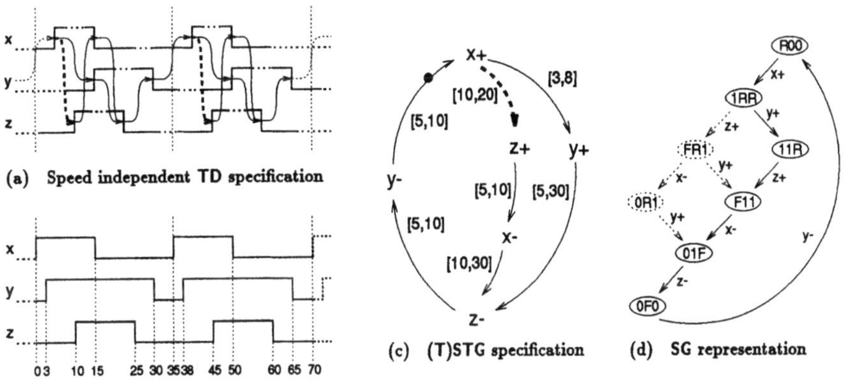

(a) Speed independent TD specification

(b) Timed TD specification (minimum firing times)

(c) (T)STG specification

(d) SG representation

Figure 1: Behavioural design specifications

In order to include additional timing information related to signal transitions, STG models could be extended [BH94][Van93][VGL94] as to specify timed interface circuits by assigning lower and upper bounds of the firing time on each arc p of an timed STG, denoted as $b(p)$ and $B(p)$, respectively. In Figure 1a,b two different timing diagrams are presented: a speed independent version corresponding to the STG without timing information on its arcs and a timed version. For a detailed description of scheduling and firing transitions by including timing intervals in STG specifications, denoted as TSTGs, we refer to [VGD92][BH94].

A state graph (SG) $\Phi = \langle M_S, T, \delta, m_0 \rangle$ may be derived from Σ by performing a reachability analysis according to the Petri Net firing rule [Chu87]. The transformation of m into m' by firing an enabled transition t, is denoted as $m[t\rangle m' \Leftrightarrow m' = \delta(m, t)$ with the partial function δ: $M_S \times T \rightarrow M_S$. $M_S = \{m : S \rightarrow \{0, 1, R, F\}\}$ denotes the set of all states, where each vector $\langle m(1), m(2), ..., m(n) \rangle$ specifies the value of each signal $s_i \in S$ of the system when the system is in that state. The initial state is denoted as m_0. Figure 1d shows the complete SG representation of the STG specification outlined in Figure 1c without including the timing information on the STG arcs. The reduced SG derived form the timed STG (TSTG) is given by removing the dotted states and arcs from the SG in Figure 1d. For more information on deriving reduced logic implementations from timed input specifications we refer to, e.g., [MM93].

Specification of delay insensitive circuit behaviour may be denoted in a completely different way to the outlined STG method. First, language–based specifications such as communicating sequential processes (CSPs) [Mar90], and second by means of asynchronous finite state machines (AFSMs). CSPs seem to be well–suited for the representation of large designs on a high abstraction level but problems may arise when attempting to specify concurrency within a process. Asynchronous finite state machines [CL93][YDN93] have already been successfully used for hazard-free synthesis of asynchronous controllers, but neither CSPs nor AFSMs incorporate timing constraints as it is usually required for a consideration of wire and environment delays [HM90][MM93][BH93].

3 The CAE framework GRACE, an overview

In the following section we give a description of our CAE framework for asynchronous circuit synthesis. The Graphical Asynchronous Circuits Environment (GRACE) consists of five main components as depicted in Figure 2.

The object-oriented data administration is central to all components. One important reason for using an object-oriented representation is the introduction of independent layers of object classes. Therefore, we separate the universal data structures, which include new object-oriented abstract graph structures as well as powerful C++ class libraries for X-Window based graphical representations, and the problem specific data objects like, e.g., STG or state graph (SG) representations. A more detailed description of the general object-oriented concepts may be found elsewhere [BBH94].

The OSF/Motif-based graphical front-end consists of the powerful STG editor SignEd and of the wavefront postprocessing tool TimEd that handles the visualization of the equivalent timing diagram while editing a protocol specification in form of an STG. Further on, TimEd visualizes the results from exercising timing analysis algorithms in order to calculate the schedule of all signal transitions of an STG specification, as outlined in Figure 1, or to determine precedence relations between particular transitions.

The import/export tool ImEx is responsible for the communication with other STG-based synthesis packages.

These graphical front-end was implemented using the universal C++ class library proposed in [BBH94]. One of the new features of this library is object representation in Adobe's Postscript

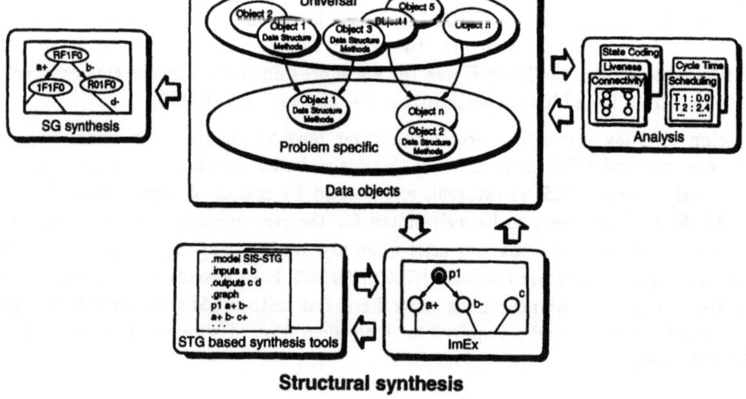

Figure 2: Components of GRACE

language. Thus, each imported STG specification and its corresponding timing diagram can not only be visualized, but also be represented in Postscript, a feature, which is important for documentation and visual validation.

The analysis component hosts a library of algorithms for the assessment of the problem specification in STG terms. Since initial specifications of application specific protocols are not necessarily hazard-free or even consistent [LKS91], powerful analysis means are mandatory to unveil several problems, such as non-feasible initial conditions or wrong schedules during protocol operation [BH94]. Finally, the synthesis component produces the structural description of the logic circuits (netlist) according to the validated STG representation of the communication protocol.

4 The graphical front-end

The graphical front-end of GRACE includes two main components: an STG editor SignEd outlined in Figure 3, and the wavefront postprocessor TimEd, as depicted in Figure 4.

4.1 SignEd

The editor SignEd represents an interactive X–Window–based graphic tool allowing for user–friendly specification of STGs in order to obtain a clearly arranged representation of the communication protocol. Some of its features are menu based command entry, user-defined configuration of input devices, and a set of visualization and manipulation commands, such as node

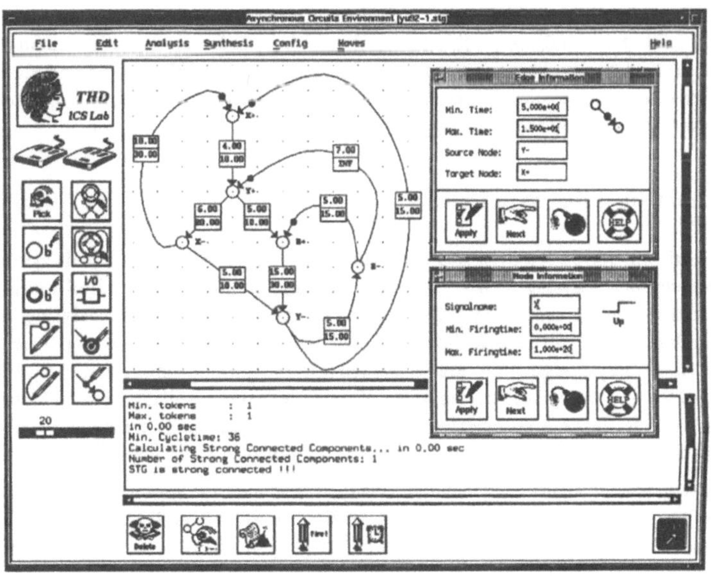

Figure 3: The editor SignEd

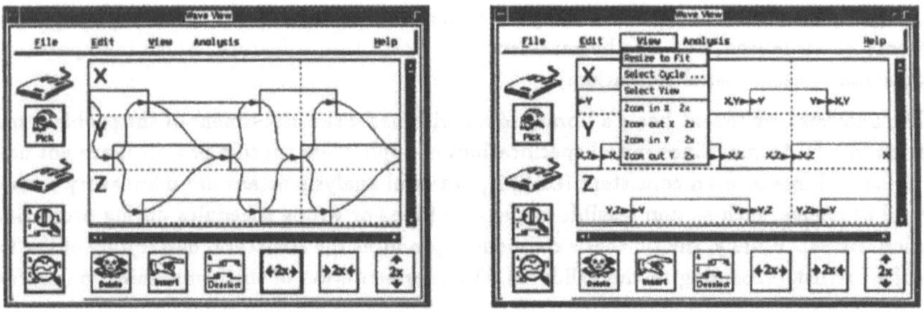

Figure 4: Selecting different views in the wavefront postprocessor TimEd

and arc editing or parameterization of objects..

SignEd is integrated into the application shell of GRACE, such that graph and net specific characteristics, like token flow and strong connected components, are first calculated by the analysis component as outlined in Section 4 and then are visualized using the SignEd back-end.

4.2 TimEd

The waveform postprocessing tool TimEd supports timed specifications of asynchronous interface circuits. Performing a minimum scheduling of the signal transitions in order to derive timing relations between any pair of transitions as well as the minimum pipelining rate of the specified circuit, TimEd supports an appropriate presentation of the results.

By using TimEd a designer is able to select signals in order to check maximum separations between particular signal transitions for an assessment of possible overlapping zones between these transitions. Separations between concurrently enabled transitions are important for, e.g., logic reduction during the synthesis process [MM93].

5 Analysis and synthesis

Many algorithms have been proposed for various stages of asynchronous circuit synthesis. General methods for analysing STG specifications corresponding to their initial marking [BH93] and their state coding properties [YS92] as well as specific methods for analysing timed STGs (e.g. pipelining rate calculation, interval based schedules, etc.) are among these proposals. Some of these analysis algorithms are integrated into the graphical front-end of GRACE in order to arrange for an appropriate presentation of possible errors resulting from STG specification. Based on this feedback a specification may be refined by an user, resulting eventually in an error-free STG representation of the communication protocol to be implemented.

Another group of algorithms includes synthesis methods for speed independent and timed specifications, respectively. In the current version of GRACE state graph synthesis and the derivation of Boolean equations is supported by built-in algorithms. The synthesis tool SIS from UCB may be used for minimization and technology mapping as well.

5.1 Analysis component

The initial marking of a strongly connected STG has to be checked in order to guarantee that every simple cycle contains more than one token and every arc belongs to at least one simple cycle that contains exactly one token. Figure 5a shows the visualization of such a check within the framework. Referring to the error message box, GRACE supports a visual browsing between the detected errors as well as an automatic refinement of the incorrect marking, while preserving this marking as far as possible. For a detailed description of the algorithms we refer to [BH93][BBH94].

Other essential conditions for deriving correct SGs are the USC (Unique State Coding) or CSC (Correct State Coding) properties [YS92][LMBS92]. In the following we give an example for a USC check performed by GRACE. An STG has the USC property if two distinct states in the SG derived from the STG do not have identical binary codes. It is clearly possible to have an algorithm that first generates an SG and then checks the SG for this property. However, working directly at STG level with a substantial lower graph complexity, a designer can more easily suggest ways to fix the problem. Different algorithms for USC are presented in the literature. In GRACE a path-oriented method based on complementary path detection is integrated. For a detailed algorithm description and more information about complementary paths we refer to [YS92]. This method has the advantage of an easier visualization of the USC property. Due to its intuitive correlation of USC to an STG specification an user may derive possible solutions to modify the STG in order to ensure the USC property.

The message box in Figure 5b shows the results of a USC check performed by GRACE on our

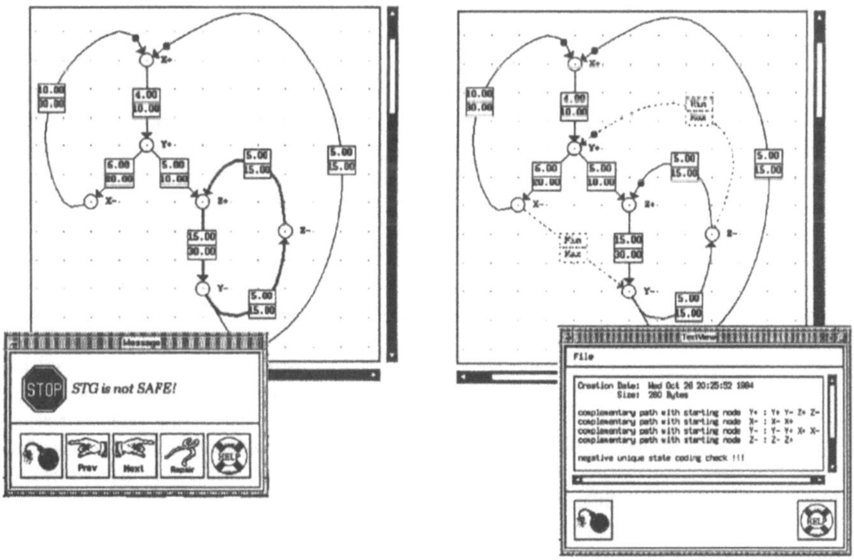

(a) Initial marking check (b) USC check

Figure 5: STG analysis

running example. The check is negative since complementary paths are detected. A possible modification of the STG in order to guarantee the USC property is outlined in the graphic window of Figure 5b.

5.2 Synthesis component

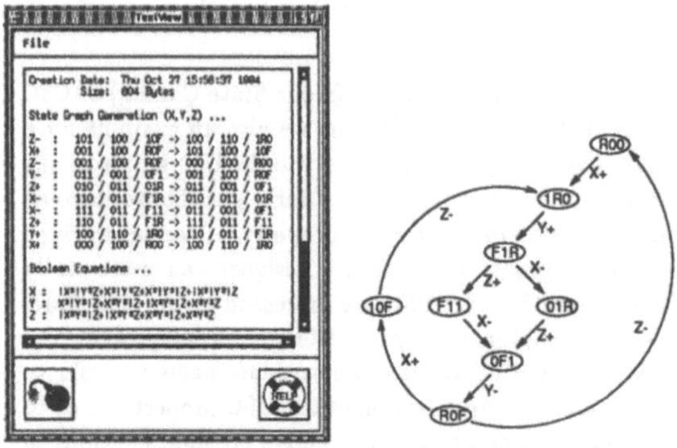

(a) (b)

Figure 6: (a) GRACE synthesis results (b) SG representation

Performing a complete state coding on the analyzed and correctly modified running example depicted in Figure 5b leads to the results given in Figure 6. In addition, the corresponding Boolean equations are generated. These equations may then the input for SIS as described in the example outlined in Section 7. New algorithms for generating reduced SGs corresponding to timed STG specifications [MM93] are also part of this component.

6 Import / export

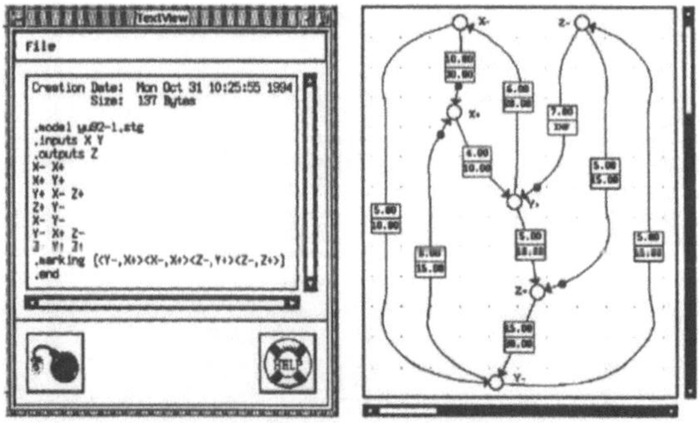

(a) (b)

Figure 7: (a) SIS format (b) Graphical representation

In this section the import / export component of GRACE, called ImEx, is presented. Using ImEx textual STG specifications generated for or by other synthesis tools may be imported into GRACE in order to provide the utilization of the enhanced graphical analysis features as outlined above. For an appropriate representation the universal object-oriented STG model [BH93] has been extended by new algorithms for constructing geometric representations from textual specifications. Therefore the vertices of the graph are assigned to different layers and then the crossings of arcs between this layers are minimized. Finally, the vertices of each layer have to be moved in order to beautify the total layout. Quadratic programming methods and heuristics have been implemented in order to solve this problem. The final graph may be visualized using polygon and spline arcs as well. For a detailed description of the algorithms and their results we refer to [Bau94]. Figure 7 shows the the import of our running example from a textual description in Berkeley SIS format.

Another component of ImEx is the export of STG models generated within GRACE to textual representations. This interface may be easily extended to utilize new STG specification formats.

7 An example

An example of a complete design flow as processed by the framework GRACE, is outlined in Figure 8. We start therefore from the behavioural description of a simple handshake circuit given in [HM90] that is, e.g., imported from SIS format, then we analyse and modify with GRACE. Finally the Boolean equations for the corresponding output signals Aout and Rout are generated. For minimization and technology mapping, e.g., SIS may be used.

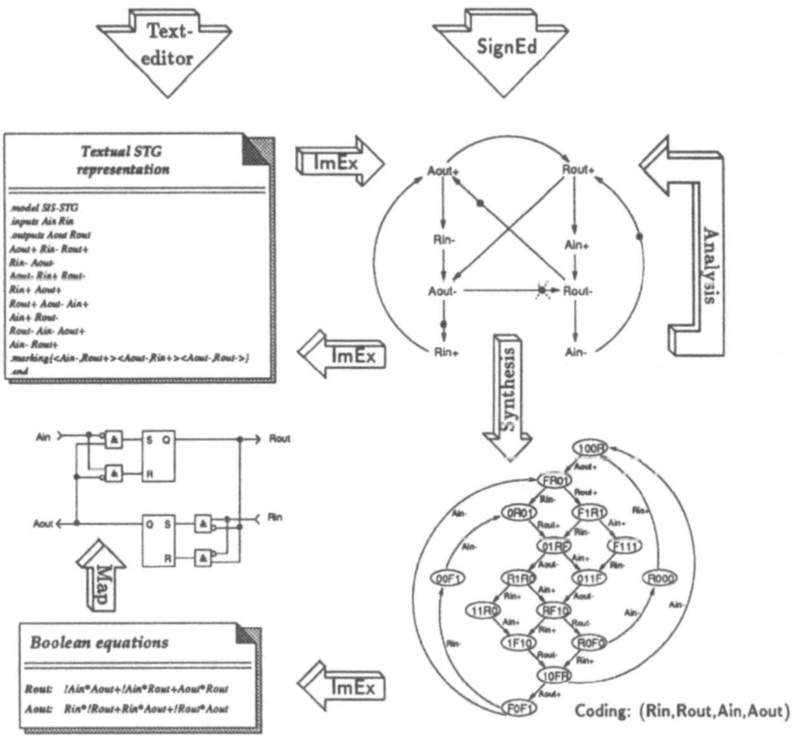

Figure 8: Design Flow

8 Conclusions

Many algorithms have been proposed for analysis and synthesis of asynchronous control circuits as outlined in this paper. However, no comprehensive visualization of the results and no complete graphical specification, analysis, and synthesis system, in which all of these algorithms are employed, has been reported to date. Therefore, we propose GRACE, a graphical asynchronous circuit environment that is both, a framework for implementing and evaluating new algorithms and a tool for automatic synthesis of asynchronous circuits using external state of the art synthesis tools by ImEx. Past experience has shown that in comparison to alphanumeric environments a designer may gain a lot of time during the specification and analysis process by using GRACE.

References

[Bau94] Jens Baumgart. Object–oriented implementation and visualisation of a graph–based data model. Master's thesis, Dept. of Computer Science, Technical University of Darmstadt, 1994.

[BBH94] Uwe Baake, Werner Bachmann, and Sorin A. Huss. High–level specification and validation of asynchronous communication protocols in an object–oriented graphical design environment. In Proc. of the International Conference on Concurrent Engineering and Electronic Design Automation, pages 349–354, April 1994.

[BH93] Uwe Baake and Sorin A. Huss. Object-oriented representation, analysis and scheduling of signal transition graphs. In IEEE Proc. of the International Symposium on Circuits and Systems, pages 2737–2740, May 1993.

[BH94] Uwe Baake and Sorin A. Huss. Scheduling of signal transition graphs under timing constraints. In IEEE Proc. of the International Symposium on Circuits and Systems, pages 205–208, June 1994.

[BLM+92] R. K. Brayton, L. Lavagno, C. Moon, R. Murgai, A. Saldanha, H. Savoj, E. M. Sentovich, K. J. Singh, P. R. Stephan, and A. Sangiovanni–Vincentelli. Sis: A System for Sequential Circuit Synthesis. Memorandum UCB/ERL M92/41, Electronics Research Laboratory, University of California, Berkeley 94720, May 1992.

[Bor89] Geatano Borriello. Synthesis of mixed synchronous/asynchronous control logic. In IEEE Proc. of the International Symposium on Circuits and Systems, pages 762–765, 1989.

[Chu87] Tam-Anh Chu. Synthesis of Self–timed VLSI Circuits from Graph–theoretic Specifications. PhD thesis, M.I.T. Laboratory for Computer Science, June 1987.

[CL93] T.-A. Chu and C. K. C. Leung. An efficient critical race–free state assignment technique for asynchronous finite state machines. In ACM/IEEE Proc. of the Design Automation Conference, pages 2–6, 1993.

[HM90] Andy Hung and Teresa H.-Y. Meng. Asynchronous self–timed circuit synthesis with timing constraints. In IEEE Proc. of the International Symposium on Circuits and Systems, pages 1126–1130, 1990.

[LKS91] L. Lavagno, K. Keutser, and A. Sangiovanni–Vincentelli. Algorithms for synthesis of hasard free asynchronous circuits. In ACM/IEEE Proc. of the Design Automation Conference, pages 302–308, 1991.

[LL91] Kuan-Jen Lin and Chen-Shang Lin. Automatic synthesis of asynchronous circuits. In ACM/IEEE Proc. of the Design Automation Conference, pages 296–301, 1991.

[LMBS92] Luciano Lavagno, Cho W. Moon, Robert K. Brayton, and A. Sangiovanni–Vincentelli. Solving the state assignment problem for signal transition graphs. In ACM/IEEE Proc. of the Design Automation Conference, pages 568–572, 1992.

[LS93] L. Lavagno and A. Sangiovanni-Vincentelli. Algorithms for Synthesis and Testing of Asynchronous Circuits. Kluwer Academic Publishers, 1993.

[Mar90] Alain J. Martin. Formal program transformations for vlsi circuit synthesis. In Edsger W. Dijkstra, editor, Formal Development of Programs and Proofs. Addison Wesley, 1990.

[Men90] Teresa H. Meng. Synchronization Design for Digital Systems. Kluwer Academic Publishers, 1990.

[MM93] Chris J. Myers and Teresa H.-Y. Meng. Synthesis of timed asynchronous circuits. IEEE Transactions on VLSI Systems, 1(2):106–119, June 1993.

[MM94] Teresa H. Meng and S. Malik. Asynchronous Circuit Design for VLSI Signal Processing. Kluwer Academic Publishers, 1994.

[RY85] L. Y. Rosenblum and A. V. Yakolev. Signal graphs: From self-timed to timed ones. In Proc. of the International Workshop on Timed Petri Nets, pages 199–206, 1985.

[Van93] Peter Vanbekbergen. Synthesis of Asynchronous Controllers from Graph–theoretic Specifications. PhD thesis, Katholieke Universiteit Leuven, September 1993.

[VGD92] P. Vanbekbergen, G. Goossens, and H. De Man. Specification and analysis of timing constraints in signal transition graphs. In IEEE Proc. of the European Design Automation Conference, pages 302–306, 1992.

[VGL94] P. Vanbekbergen, G. Goossens, and B. Lin. Modeling and synthesis of timed asynchronous circuits. In IEEE Proc. of the European Design Automation Conference, 1994.

[YDN93] Kenneth Y. Yun, David L. Dill, and Steven M. Nowick. Practical generalisations of asynchronous state machines. In IEEE Proc. of the European Design Automation Conference, 1993.

[YS92] Meng-Lin Yu and P. A. Subrahmanyam. A new approach for checking the unique state coding property of signal transition graphs. In IEEE Proc. of the European Design Automation Conference, pages 312–321, 1992.

[YVL93] C. Ykman, P. Vanbekbergen, and B. Lin. Alpha release of the ASSASSIN asynchronous compiler. Technical report, IMEC, 1993.

PART EIGHT

Standards

21

Design Representation in EDIF Version 3 0 0 and CFI Version 1.0

Hilary J. Kahn
Department of Computer Science
University of Manchester
Manchester M13 9PL, UK
e-mail: hkahn@cs.man.ac.uk

Abstract

The development of frameworks and integrated CAD solutions has, over the past ten years or so, led to the development of a number of standards which play a key role in the electronic product design and development cycle. Prominent among these are the Electronic Design Interchange Format (EDIF) and, more recently, the CAD Framework Initiative Design Representation (CFI DR). These two standards has been characterised by the use of a formal specification as an integral part of their development process. In both cases the published standard includes a specification in the form of an EXPRESS Information Model. CFI DR and EDIF also share another characteristic; they both cover the domain of electrical connectivity representation. In the case of CFI DR 1.0, connectivity is the main design aspect covered. EDIF Version 3 0 0 has a wider scope, including schematics as well as connectivity. In this paper, the information models of both CFI 1.0 and EDIF Version 3 0 0 are reviewed. The degree of overlap that exists between them is examined and indications given of a common or 'core' information model from which future versions of CFI and EDIF may be derived.

1. Introduction

It has been clear for a number of years that there is value, both from a theoretical standpoint and from a pragmatic point of view, in working towards the alignment of the underlying 'models' used by EDIF [1] and CFI DR [2]. Both standards relate to a fundamental part of the infrastructure of EDA systems. EDIF deals with the transfer of design information between EDA systems or from design system to manufacturing environment; CFI provides closely coupled communication between CAD tools and their support environments. Neither standard is targeted at end users, and both are independent of specific design flows or design stages.

Technical interchange of ideas has been fostered over the years by, for example, having a CFI representative on the EDIF Technical Committee during the development of EDIF Version 3 0 0 and by EDIF experts being involved in the review of the CFI DR information model. However, the climate has, until recently, not been right for the next step, the consideration of a formal technical alignment of EDIF and CFI DR. The catalyst for change, and hence for the study whose findings are reported here, was the decision by the EDA Companies consortium (EDAC) earlier in 1994 that they wished the various standards bodies and standards creators to make some positive move towards alignment. This strong push from the industry comes, of course, from the entirely pragmatic view point that it is not cost effective for the major EDA vendors to be involved in the development of a number of, possibly competing, standards. The cost is high in terms of money and manpower to maintain competence in different standards, and in terms of development and support effort to build compliant software.

As a result of the EDAC initiative, a series of technical meetings took place between Laurence Brevard of CFI and Hilary Kahn representing EDIF. The initial goal was to investigate the degree of overlap between EDIF Version 3 0 0 and CFI DR 1.0 and to advise the EDA companies whether or not a core or common model between the two standards was feasible. In the event, the result was not only the identification of areas of overlap but also an assessment of some of the ways in which maximal mappings between EDIF and CFI could be achieved and a proposal for the constituents of a core model. The inputs into this assessment process were the two information models and a technical report [4] which presents in great detail a review of the minimal mappings that are supported by the two basic models.

2. The EDIF and CFI Models

The information models of EDIF Version 3 0 0 and CFI DR both use EXPRESS Version N14 [3]. The use of the same modelling language in many ways eases the task of comparing the models. However, each model is written in a specific style which may, in part at least, be justified by the subtle differences in the EDA system domains to which they apply i.e. Programming Interface (PI) vs. file-based design transfer

2.1 CFI DR 1.0

The CFI 1.0 standard, published in 1992, includes an information model of design representation as part of the formal documentation. The model is small, consisting of seven schemas with a total of just over 20 entities (domain objects). The schema structure reflects the objects in the domain rather than broader aspects of the design domain, Hence, the schemas relate to: library, cell, net, port, instance and portInstance, as well as a unifying 'basic object model' schema. The goal of the model is the formalised specification of the domain covered by CFI DR; this is used as the basis for the PI function specifications which are the real 'product' in the CFI DR standard. CFI supports the representation of non-parameterised design.

The CFI DR model has a consistent style in terms of the representation of basic characteristics. For example, the names of entities have been chosen to fit in with the intended PI notation. So, in an EXPRESS class structure (supertype-subtype), the name of the supertype typically is the most commonly used term; the subtypes are then appended with additional characterising names. An example is illustrated in the model fragment in Figure 1.

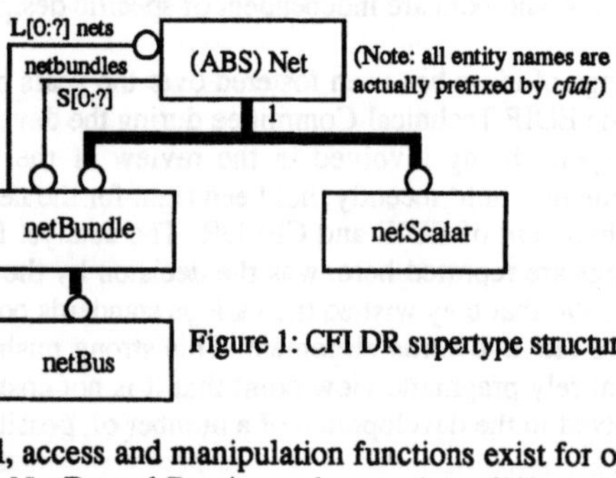

Figure 1: CFI DR supertype structure

In the PI based on this model, access and manipulation functions exist for operating on the net object itself (e.g. cfidrNetDetachPort) and on its different subtypes (e.g. cfidrNetBundleInsertNet).

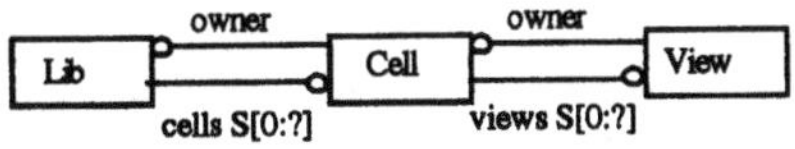

Figure 2: CFI owner attribute

Another style used in the CFI DR model is the explicit identification of 'owner' relationships. Although EXPRESS provides mechanisms which establish existence dependency (i.e. the INVERSE attribute), in the CFI DR model this is not used. Instead an explicit owner is defined as illustrated in Figure 2.

This mechanism is justified by the importance of the role played by the owner relationship in a PI. In CFI DR, there are a number of functions which explicitly manipulate the owner attribute (e.g. cfidrCellGetOwner, cfidrPortSetOwnerView) and the create functions (e.g. cfidrPortBundleCreatePortBundle) all have the owner attribute as a parameter. It should be noted that, although it provides the basis for the PI in terms of objects and relationships which must be supported, the CFI information model does not dictate the syntactic form of the CFI DR PI. That is determined by considerations of issues such as run-time efficiency and ease of use.

2.2 EDIF Version 3 0 0

The EDIF information model is much larger than that of CFI, mainly because it covers a wider area of design representation, including schematic diagrams and design configuration, as well as connectivity. In addition, as EDIF deals in static and in parameterised design representation, it is defined at two separate levels. EDIF Level 0 handles all the requirements for the transfer of non-parameterised designs and libraries; EDIF Level 1 adds to this the capability to represent frames and parameterised designs. Hence, the formal standard includes two separate information models; one for Level 0, the other for Level 1. For the purposes of this discussion, only the Level 0 information model will be considered.

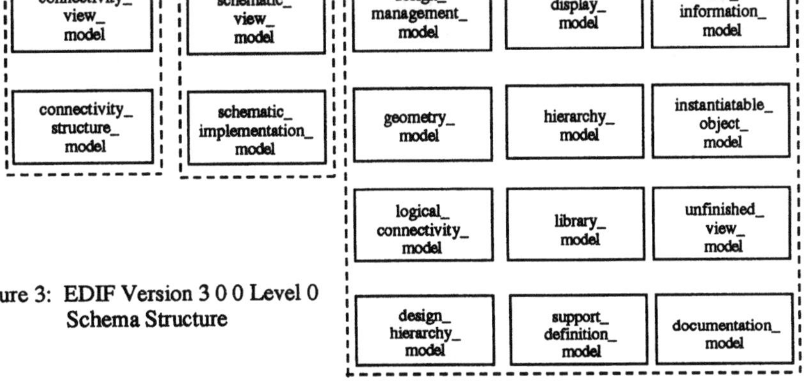

Figure 3: EDIF Version 3 0 0 Level 0 Schema Structure

The Level 0 information model consists of 16 schemas divided into separate sub-domains. Some of the schemas are specific to the type of EDIF view being described; the majority of the schemas, however, cover information which is either common to more than one type of view e.g. the hierarchy_model schema, the logical_connectivity_model schema, or represents fundamental concepts e.g. the geometry_model schema.

Within the EDIF information model care has been taken to use consistent object naming conventions; these are typically directly reflected in the corresponding EDIF keywords in the syntax. However, just as the CFI information model does not dictate the form of the CFI DR PI calls, so the EDIF information model does not determine syntactic issues such as the underlying use of 'define before use' in the EDIF format or the Lisp-like lexical structure.

However, some fundamental relationships specified in the model are found reflected in the EDIF syntax. These include the concepts of 'containment' and 'uniqueness' in a given context. In EDIF, the relationship that establishes the existence dependency of an object on another object, is not established using the EXPRESS INVERSE construct; instead, the function 'belongs_to_one' is used. Here it is illustrated used in the context of a cell belonging to one library. Note also that the cells in a library are defined as a set.

```
ENTITY library;
        cells:     OPTIONAL SET [1:?] OF cell;
        ...
END_ENTITY;

ENTITY cell;
        ...
WHERE
containment_constraint :   (* A cell belongs to one library. *)
        belongs_to_one (SELF, 'library_model..library.cells');
END_ENTITY;

FUNCTION belongs_to_one
        (entity_instance : GENERIC; role :  STRING) : BOOLEAN;
        RETURN(SIZEOF(USEDIN(entity_instance, role)) = 1);
END_FUNCTION;
```

3. Design Hierarchy

Within EDA systems design information is typically held in some form of hierarchy. Where information is held in libraries, the natural form of the hierarchy is a folded definition hierarchy. In other words, each object (typically a 'cell') in the library is the definition of some design object. It is a two-level hierarchy description in which the object is defined strictly in terms of a network of instances of simpler objects. The hierarchy is always two level, regardless of whether the instanced objects are themselves hierarchical. See Figure 4.

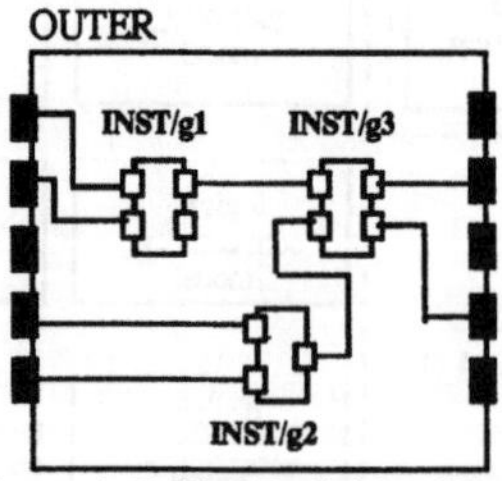

Figure 4. Folded Definition Hierarchy

In both CFI DR and EDIF, this folded definition hierarchy is used for library object (e.g. cell) definitions. In EDIF, objects may be given properties and attributes when they are defined; those attributes and properties may be inherited when the objects are themselves used through the instantiation mechanism. In addition, EDIF provides for the overriding of the inherited attributes and properties at the point of instantiation. The inheritance and overriding of properties and attributes is not supported in CFI DR.

A complete design described in an EDA system may additionally be represented (possibly temporarily during processing) as an unfolded or expanded occurrence hierarchy. Where an unfolded occurrence hierarchy is not supported, a flat representation may be used instead. These forms of design occurrence representation are shown in Figure 5. They are not directly supported in CFI. However, in EDIF, there is a mechanism for configuring an unfolded design hierarchy and for annotating property and attribute values to the occurrences.

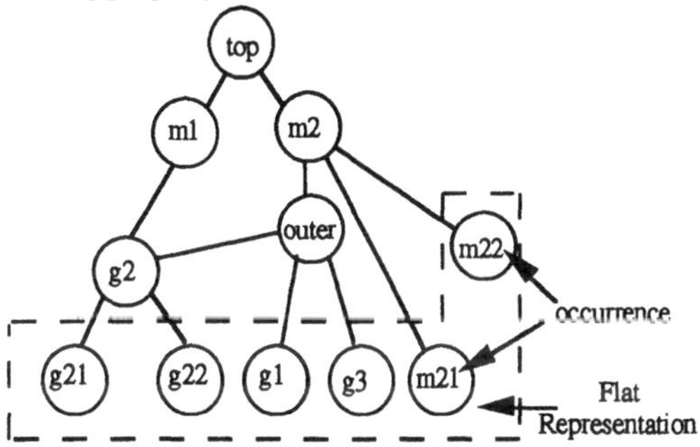

Figure 5: Unfolded Occurrence Hierarchy and Flat Representation

4. The Goal of a Common Model

Unless active steps are taken to avoid divergence, the development process for both EDIF and CFI will evolve as shown in Figure 6 (a). Each standard will resolve issues strictly within its own purview without regard to future coherence.

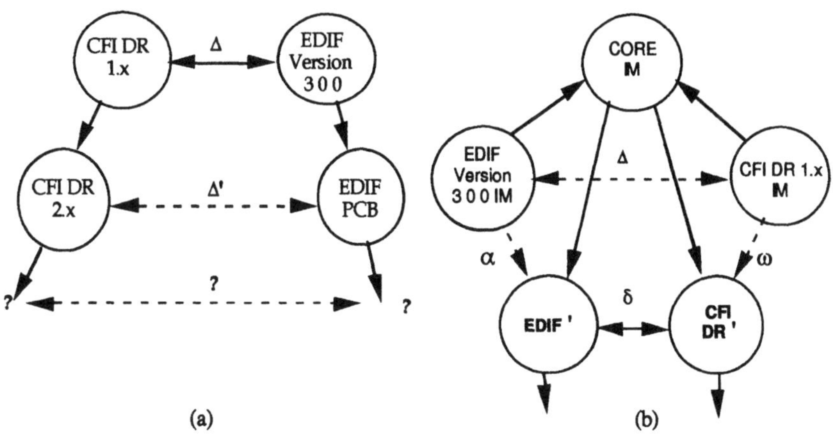

Figure 6: (a) The Current Situation (b) The goal of the EDIF-CFI alignment activity

What is being sought by the EDA industry is a single underlying model from which the standards that they need to support have been derived. This is shown in Figure 6 (b), where the core information model is obtained by a reverse engineering process from the present EDIF and CFI models and then itself used to create future EDIF and CFI versions. The primary aim is to reduce the divergence from Δ to δ. The intent is that δ should be minimised - and that, at the same time, α and ω should be kept as small as possible. This process encourages the future standards developments to take place in step with each other.

5. Domains of Overlap

The scope of EDIF Version 3 0 0 is connectivity, schematics and design configuration, together with the support of libraries and definition hierarchy; that of CFI DR is connectivity, libraries and definition hierarchy. There is clearly therefore a mismatch between the current coverage of the two standards. Should CFI DR find it appropriate to move towards coverage of those domains outside its present scope, it would seem natural that EDIF Version 3 0 0 would be a major resource from the information model point of view. In the area where overlap exists, that of connectivity and hierarchy, a deeper analysis is required. The following sections review some of the relevant aspects.

5.1 Hierarchy

In EDIF, the hierarchy is defined in terms of **library → cell → cluster → view**; in CFI DR only **library → cell → view** is used (as in EDIF Version 2 0 0). Furthermore, in EDIF, the object that is instantiated is the cluster whereas in CFI DR, it is the view. The issues therefore become: why did EDIF introduce the concept of cluster and is cluster a relevant concept in a core model?

The cluster introduced into EDIF allows all the cell representations (views, symbols) that share a common interface to be grouped. During any instantiation process a 'contract' is established between the instantiating environment and the instantiated object. In design terms, what is required is that the instantiating environment can relate to the interface (i.e. the ports) of the instantiated object. The local representation used to show this instantiation relationship in an EDA system may be the drawing of a symbol on the instantiating schematic page or a reference to the object from a language context. However, the true representation that will be used when the design hierarchy is traversed to produce the occurrence tree for a target process (such as simulation or the manufacture of the design) is selected by the requirements of that target process. This is supported in EDIF by instantiating a cluster and then allowing the design configuration process to select a suitable view from among those belonging to the given cluster. No equivalent mechanism exists in CFI DR although CFI does currently offer a view select mechanism which allows the describer of the selected view to be changed - but without any constraints on the suitability of the new view or its interface.

The characteristics of a cluster as described above are relevant beyond the domain of EDIF. It is therefore appropriate for the cluster to become a core model concept. Note that a benefit of cluster as a grouping mechanism for cell representations that have a common interface is that it is a step towards simplifying mappings between ports of different views.

5.2 Connectivity: Logical and Structural

In EDIF, a distinction is made between **logical** connectivity and **structural** connectivity. Logical connectivity captures abstract view-wide connectivity in terms of single-bit objects (which are called signals) linking single-bit ports. Any type of view which supports connectivity data includes logical connectivity. The logical connectivity is the basis for the structural connectivity which 'implements' it. Structural connectivity is expressed in terms of nets and busses, which may themselves be structured into sub-nets and sub-busses. Structuring into sub-nets and sub-busses allows implementation attributes, such as timing data and routing criticality, to be associated with parts of connections and point-to-point links. CFI DR currently supports the concepts of logical connectivity, but not of structuring.

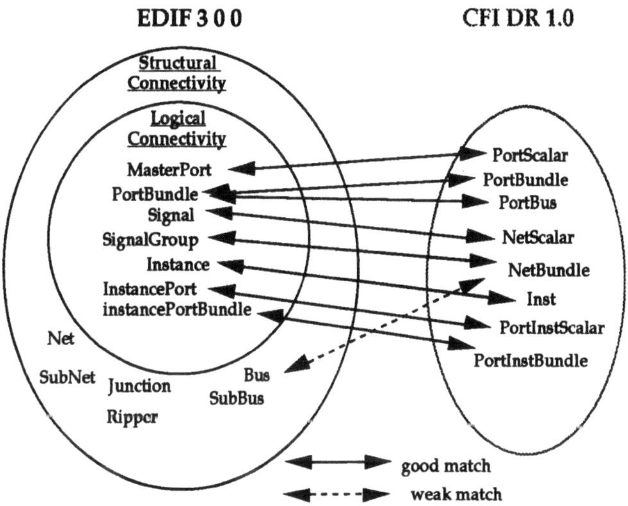

Figure 7: EDIF and CFI Connectivity Overlap

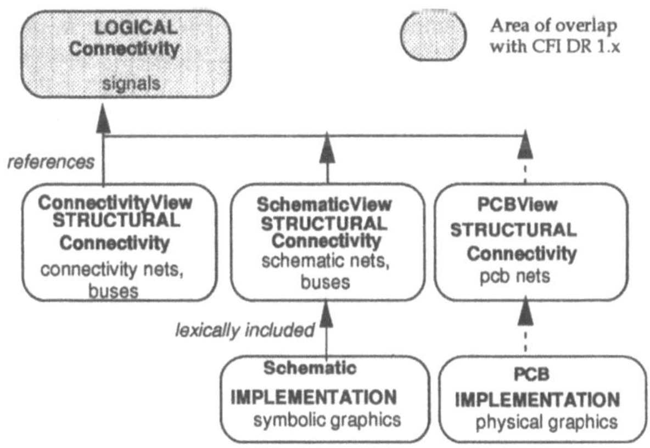

Figure 8: Relationship between Logical and Structural Connectivity in EDIF

In EDIF and CFI DR the fundamental connectivity supported in both models in strictly single-bit. In EDIF signals join ports; in CFI DR netScalars join portScalars. Both standards superimpose on this fundamental connectivity model facilities for representing wide connectivity. In EDIF, busses implement groups of signals (signalGroups) and are associated with 'wide' ports (portBundles); in CFI DR, netBundles group netScalars and are associated with portBundles. In neither case is connectivity actually established by the bus/bundle mechanism. In EDIF the existence of the wide connectivity objects is required to match the models commonly used in EDA systems and to support, for example, the drawing of bussed-objects on a schematic; in CFI DR, the wide objects provide a commonly used short-hand mechanism for manipulating grouped objects.

Figure 8 shows the general architecture used in EDIF to relate logical and structural connectivity. This approach supports extensibility of the domains covered, as is illustrated by current developments which are enhancing EDIF to include a PCB/MCM capability. These concepts are relevant to an EDIF-CFI core model, not only because they are already fundamental to EDIF, but because CFI is moving into new domains, including the PCB area and the incorporation of timing data.

5.3 Naming Philosophy

The study of the EDIF and CFI DR information models and reference manuals highlighted some interesting points about how names associated with objects are used in each standard. In order to understand what is required, it is necessary to recognise the purposes to which these object names may be put. Objects can be named in order to make them identifiable in a given context (i.e. they have a unique name) and object names are used as 'pointers' to allow objects to be referenced. In EDA systems, objects are also given names which the user selects - these user-defined strings allow the designer to annotate a design in a manner meaningful to him; in this context, name aliasing is also commonly found. Finally, objects may be given default names by the EDA system itself- a function that is needed where the user elects not to provide a name.

In EDIF Version 3 0 0, a move was made away from the identifier and name overloading of EDIF Version 2 0 0. As a result, the **EDIF identifier** is used exclusively for making an object unique in a given context and for referencing within the EDIF file. Support for user-defined names and for names supplied automatically by the EDA system is provided by a new **nameInformation** construct. In CFI, **cfidrNamedObjectName** establishes object uniqueness but object referencing is done using the **cfidrObjectId** (OID) which is the reference handle unique to a given CFI session. As with EDIF Version 2 0 0, CFI provides no mechanism for capturing the user- or system-supplied names for an object although in some cases (see 5.4.1), overloading of the cfidrNamedObjectName mechanism is used to achieve a similar effect.

For a core model there is no need to define a single implementation mechanism for specifying object uniqueness in a context/session; CFI will continue to use the OID and EDIF will use the EDIF identifier. However, a mechanism for capturing the 'real' object name is required; a solution based on the EDIF **nameInformation** is suggested.

5.4 Grouping Philosophy

One of the few areas where EDIF and CFI DR deal with a single issue in different ways is that of grouping objects. A number of different types of object may be grouped: ports, signals, nets, instances. The reasons for forming these groupings are not always consistent. The grouping of ports and signals/nets are considered here.

5.4.1 Grouping Ports.

In EDIF, portBundles are formed by grouping existing ports and portBundles. PortBundles are defined in the cluster interface. A port/portBundle may be included in more than one portBundle but a port or portBundle can appear at most once in a given portBundle.

LocalPortGroups in EDIF are handled in a similar manner to portBundles. In CFI DR, cfidrPortBundles provide a defining context for the cfidrPorts and cfidrPortBundles they contain. Hence a cfidrPort/PortBundle can be in at most one cfidrPortBundle.

Despite the apparent difference in the portBundle in each standard, the fundamental fact remains that in both standards the real connectivity objects are the single-bit ports and their single-bit connections, as noted above. It is therefore necessary to examine what each standard gains from the mechanism adopted. The EDIF model directly reflects the fact that the ports are the atomic objects and portBundles are convenient groupings. In CFI, the fact that a portBundle defines the ports it contains supports a specific requirement for the interface operation of 'connect by name'. As the cfidrPortBundle provides a name scope for the objects that it contains, a shorthand operation that causes all objects with the same 'name' to be connected is possible. Clearly, this relies on the overloading of cfidrNamedObjectName noted above.

It is therefore proposed that a compromise model be adopted as the core representation. The use of the EDIF-like mechanism for grouping existing ports into portBundles will be maintained although a constraint to ensure that a port is included in at most one portBundle may be agreed. In addition, rather than rely on the overloading of cfidrNamedObjectName to associate a name with the composite object members, a new attribute designed specifically to support member naming is likely. Provisionally, this is referred to as the localNameTag and its effect will be to give an object grouped by a portBundle a 'name' unique to its use in that context.

5.4.2 Grouping Signals and cfidrNets.

Here the difference in the grouping mechanisms used in EDIF and CFI is minimal. In both cases, the grouping construct (signalGroup or cfidrNetBundle) groups existing single bit or grouped objects. In both cases, too, a signal/cfidrNet may appear in more than one group and more than once in a given group.

The mechanism suggested for the core model is similar to (but less constrained than) that proposed for ports above. In other words. the grouping entity will group existing objects; in addition, objects inside a group may optionally have an associated localNameTag attribute to support membership naming.

5.4.3 CFI Bus Objects

In CFI DR there are objects called cfidrNetBus and cfidrPortBus which are formally subtypes of the cfidrNetBundle and cfidrPortBundle respectively. These objects have a relatively weak definition in CFI DR. Their main distinguishing feature is that they can be treated as indexed objects i.e. their members may be addressed by index. EDIF has no equivalent subtype for signalGroup or portBundle, but it does capture knowledge of the bus-indexing mechanism used in an EDA system through the nameInformation construct. On the assumption that the core model will adopt the EDIF-like nameInformation mechanism as suggested above, it is likely that a constrained form of the mechanism already in place in nameInformation will be used in the core model to provide CFI with the equivalent of the cfidrNetBus and cfidrPortBus

6. Summary

This paper has addressed some of the issues that arose as a result of the detailed study of the EDIF Version 3 0 0 and CFI DR 1.0 information models. Other topics which the study reviewed concerned the handling of properties and attributes (where EDIF supports inheritance through the instantiation mechanism and currently CFI DR does not), the representation of global signals/ports (e.g. clocks, power), the relevance or otherwise of the EDIF external library concept and issues related to the handling of strings (where EDIF supports a wide range of options through the ISO standards). Another topic that was reviewed concerns the CFI DR 2.0 plans to support an attachment operation. This is a means of attaching a cell (view) to a library (cell) other than the one in which it was initially defined. The mechanism proposed for CFI is very similar to the UNIX™ symbolic link. No definitive model was agreed for handling this in the core model and in a future version of EDIF, but it is clear that there are a number of possible ways to satisfy this requirement.

7. Conclusions

The real result of the study is that there are no technical 'show-stoppers' that prevent EDIF and CFI from coming closer together and sharing a common information model. The review has shown that the new ideas introduced into EDIF Version 3 0 0 as a result of the experience gained through the use of the EDIF Version 2 0 0 are fundamentally sound and are able to meet many of the needs of CFI for the future. The technical evidence indicates that the core model would be largely based on the EDIF Version 3 0 0 information model, possibly with some minor adjustments.

However, whether or not the step of creating the common core model will actually be taken depends less on technical issues than on practical ones of man power investment and of commercial and political will. The EDAC Tiger Team generally found the proposals acceptable so the issue now is how to effect the CFI transition to a common model. In practice, the move will probably be phased via an intermediate stage where the new core model is supported together with mappings to the existing CFI model via 'helper' functions. For EDIF, little or no change is required to meet the constraints of a common model. Minor alterations, particularly those not involving changes to the current syntax, would probably be acceptable even for EDIF Version 3 0 0 implementations if they were agreed quickly enough. Other changes, if agreed, could be included in the future PCB/MCM version of EDIF planned for release in 1995.

8. References

[1] Electronic Industries Association. Electronic Design Interchange Format Version 3 0 0, EIA-618 December 1993 (4 volumes)

[2] CAD Framework Initiative Inc. CFI 1.0 Design Representation Programming Interface 1992

[3] ISO TC184/SC4/WG5 "EXPRESS Language Reference Manual" ISO 10303: Part 11 Version N14 April 1991

[4] Alan Williams, "Mapping between CFI 1.0 and EDIF Version 3 0 0", Version 0.1 Internal Report, CAD Group, Department of Computer Science, University of Manchester, UK, 1994

9. Acknowledgements

First of all, I would like to recognise the contribution made by Laurence Brevard of CFI to the work described here. We both learned a great deal about EDIF and CFI during our meetings and Laurence ensured that we maintained a focus on practical issues and drew on the expertise of the many people involved in creating and implementing EDIF and CFI. The interpretations presented here are, I believe, those that Laurence and I share. Thanks are also due to Alan Williams, Rachel Lau and Nigel Whitaker of the University of Manchester and Robin La Fontaine (Monsell) for advice and comments and to Don Cottrell (CFI), Rich Goldman (Synopsys and EDAC), Jeri Crowley (TI and EDIF) and Ron Christopher (IBM) for making sure that Laurence and I actually held the meetings. The financial support of the CEC through the ESIP Project (ESPRIT 8370) is gratefully acknowledged.

22

STEP Technology for ECAD Databases

Frank Buijs*
Cadlab
Bahnhofstr. 32
33102 Paderborn
Germany

Wolfgang Käfer†
Daimler-Benz F3P
Postfach 2360
89013 Ulm
Germany

Topics: *Common Databases, Standardization Activities, Exchange Formats*

1 Introduction

Standards are increasingly important for ECAD frameworks. The emerging international standard for the exchange of product model data (CAD data, CAM data, ...) is ISO 10303 "Industrial automation systems and integration - Product data representation and exchange", better known as STEP (STandard for Exchange of Product data). This paper addresses the STEP technology for ECAD databases.

STEP provides standardized data schemas, called *Application Protocols (APs)*, for a number of application domains. E.g., AP210 is a standardized data schema for "Printed circuit assembly product design data". For defining the data schemas, STEP provides the information modeling language EXPRESS. Thus, the APs are annotated EXPRESS schemas.

Part 21 of STEP defines how an ASCII file format is derived from an EXPRESS schema. Thus, an AP together with Part 21 defines a standardized file format for data exchange. E.g., AP210 together with Part 21 provides a functionality similar to EDIF-PCB.

Part 22 of STEP defines the database interface SDAI (Standard Data Access Interface), which provides a standardized access to data that conform to an AP. E.g., AP210 together with Part 22 defines a standardized database interface to PCB data. In general, applications can be implemented on the basis of the SDAI as soon as an EXPRESS schema of the data is available.

Therefore, we believe that SDAI will also be the leading standard interface for ECAD

*E-mail: Buijs@Cadlab.DE
†E-mail: Wolfgang.Kaefer@dbag.ulm.DaimlerBenz.COM

databases. However, the definition of the SDAI is still under development and is not yet published by the ISO. The goal of this paper is to present the basic concepts and features of the SDAI. The new features that will appear in the new SDAI version are also described.

Section 2 provides some STEP background. The subsequent sections present the SDAI concepts. Section 5 gives an example of an SDAI session. Section 6 presents the features that were added to the SDAI recently.

2 STEP Background

For an overview of STEP we refer to [1, 2, 6]. The goal of this section is only to give some background information before explaining the SDAI concepts.

STEP is organized in two main blocks. The first block of documents, i.e., Parts 1 - 10 (fundamentals), Parts 11 - 20 (description methods), Parts 21 - 30 (implementation methods), and Parts 31 - 40 (conformance testing), are not associated with any specific application domain. They describe generic methods and tools which are applicable to any kind of information modeling task. For the scope of this paper Part 11 [3], the EXPRESS Language Reference Manual, and Part 22 [5], the SDAI specification, are of particular interest and discussed below.

The second block of documents employ the methods described in the first block to more and more application specific environments. Parts 200 - 1199 define the Application Protocols, like Part 210 (AP210 "Printed circuit assembly product design data"). and Part 214 (AP214 "Core data for automotive mechanical design processes"). The APs use subschemas, called *resources.* Parts 41 - 100 of STEP define the integrated resources, application-independent subschemas for kinds of data that are frequently used. They are used by several APs. For example, the subschema "Product structure configuration" (Part 44) is used by both AP210 and AP214. Parts 100 - 199 define the application resources, subschemas that are more application specific. Parts 1200 - 2199 define the abstract test suites corresponding to the APs 200 - 1199.

2.1 EXPRESS

Part 11 of STEP defines the information modeling language EXPRESS [3]. With EXPRESS, and its graphical representation EXPRESS-G, data *schemas* can be defined.

An EXPRESS schema basically defines a set of entities. An entity is defined by an identifier and a set of attributes. The data type of an attribute may be a simple type (number, real, integer, binary, boolean, logical, string), an enumeration type, a select type, or an entity type. Aggregate data types (set, bag, list, array) are also allowed. The entity type attributes are used for defining the relationships between the entities.

EXPRESS also allows for the definition of *rules* in order to state consistency conditions or restrictions on the values of attributes.

For example, Figure 1 shows a simple EXPRESS schema with two entities. The attribute `members` of `port_group` defines an $1 : n$ relationship between the entities `port_group` and `port`. The uniqueness rule `UR1` defines that the names of each instance of the entity `port` are unique.

```
SCHEMA example;

ENTITY port
     name :  STRING;
UNIQUE
     UR1:  name;
END_ENTITY;

ENTITY port_group
     members :  SET OF port;
END_ENTITY;

END_SCHEMA;
```

Figure 1: Sample EXPRESS schema

Furthermore, EXPRESS supports inheritance. Entities inherit attributes from other entities that are defined as their supertypes.

2.2 Implementation Methods

For the exchange of *entity instances*, two implementation methods are defined. Part 21 of STEP defines the format of STEP Physical Files [4]. A STEP file contains mainly a list of entity instances. Each entity instance is defined by a unique identifier (an integer preceded by a '#'), an entity identifier of the corresponding EXPRESS schema, and a list with the values of the attributes. For example, the entity instances listed in Figure 2 correspond to the EXPRESS schema in Figure 1.

```
#1=port("port1");
#2=port_group((#1,#3));
#3=port("port2");
```

Figure 2: Sample STEP Physical File

Part 22 of STEP defines the database interface SDAI [5]. The functionality of the SDAI is independent of a particular programming language. For implementing the SDAI, Parts 23 - 25 of STEP define *language bindings* (C++, C, and FORTRAN, respectively).

A language binding can be either a *late binding* or an *early binding*. A late binding is independent of a particular EXPRESS schema, an early binding is not. For example, in order to give an instance `p1` of the entity `port` the name port1, a late binding provides the generic operation `PutAttribute(p1,name,"port1")`, an early binding provides the schema-specific operation `PutAttributePortName(p1,"port1")`. In the case of a late binding, the parameters like `name` are typically evaluated at run-time. In the case of an early binding, such parameters are typically evaluated at application program compile-time. Therefore, an early binding SDAI implementation performs better than a late binding SDAI implementation.

3 Data Organization of the SDAI

The SDAI is independent of the underlying data storage technology, i.e., the data may be stored in a relational database, an object-oriented database, a file system, or a combination of them. The generic name for the data stores is *repository*. A repository may be a single database or a collection of physical files. Thus, from the physical point of view, the entity instances are grouped in repositories.

Apart from the physical view, the SDAI also distinguishes a logical view. Logically, the entity instances are grouped in *schema instances*. Each schema instance has an underlying EXPRESS schema. All entity instances in a schema instance conform to this underlying schema. Schema instances are created, modified, and deleted by the application. Several schema instances with the same underlying schema are allowed.

The schema instance is the domain for referencing. For example, in the case of the three entity instances in Figure 2, entity instance `#2` references to the entity instances `#1` and `#3`. This referencing is only allowed if `#1` and `#3` are in the same schema instance as `#2`. The schema instance is also the domain for the validation of rules. For example, suppose a schema instance has the schema in Figure 1 as underlying schema. Uniqueness rule `UR1` defines that each instance of the entity `port` must have a unique name. A validation of this uniqueness rule tests if all instances of `port` in the schema instance have unique names. Note, that the domain for referencing and rule validation is a schema instance and not a repository, i.e. in our example, a validation of rule `UR1` does not test if all instances of `port` in the repository have unique names.

Within the repositories and the schema instances, the entity instances are grouped in *sdai models*. An sdai model may contain some grouping of data which has a meaning as a whole, such as a particular part description along with its geometry. A schema instance can be distributed across several repositories, but all the entity instances in an sdai model are allocated in *one* repository. Thus, each entity instance is in one sdai model, and each sdai model is in one repository.

Figure 3 illustrates the notions of repository, schema instance, and sdai model. This figure also illustrates that an entity instance in one sdai model may refer to an entity instance in another sdai model, even if the sdai models are in different repositories — provided the sdai models belong to the same schema instance.

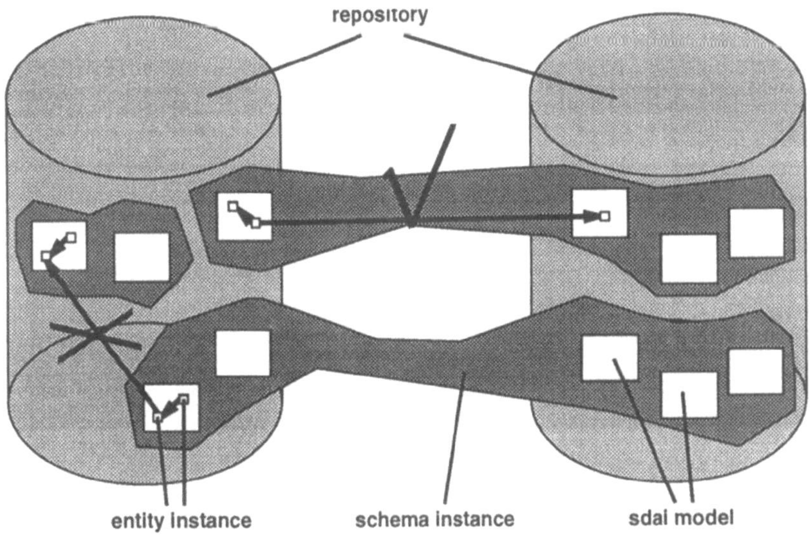

Figure 3: Logical and physical data organization

Apart from the sdai models containing application data, the SDAI also has some special sdai models containing the so-called *meta data*. These sdai models contain the session data, such as the error log and information on the data organization, and the dictionary data. The dictionary data describe the EXPRESS schemas supported by the SDAI implementation. In the case of a late binding SDAI implementation, the dictionary data are used for mapping the SDAI operations onto database operations. Since the meta data are stored in sdai models, just like the application data, the meta data can be accessed using the same SDAI operations as used for retrieving application data. The applications have read-only access to the meta data.

4 SDAI Operations

The SDAI provides a large number of operations to access the data. This section gives an introduction to the SDAI operations.

The *schema instance operations* create and delete schema instances, and validate rules. E.g., the operation `ValidateUniquenessRule(SI1,port,UR1)` tests if all instances of `port` in schema instance `SI1` have unique names. Note, that validations are not performed automatically by the SDAI, but only on demand of the application. Even if an application adds an element to a set, the SDAI does not test whether this element is already a member of the set. The SDAI merely provides an operation that tests whether the members of a set are unique.

The ***sdai model operations*** create and delete sdai models, and initiate the access to an sdai model. E.g., the operation `StartReadWriteAccessMode(M1)` opens the sdai model `M1` for read-write access.

The *entity instance operations* are used for creating, copying, modifying, and deleting entity instances. Typical entity instance operations are CreateEntityInstance, PutAttribute, GetAttribute, etc.

For the handling of aggregates, like sets and lists, iterators are used. An *iterator* is a mechanism that allows for traversing the contents of an instance of an aggregate. E.g., in the case of an instance of the entity `port_group`, the operation CreateIterator can be applied to its attribute `members` to obtain an iterator on this aggregate instance. Subsequent operations such as Next, GetCurrentMember, and RemoveCurrentMember are applied to this iterator for traversing and accessing the aggregate instance. An application may use multiple iterators on an aggregate instance. However, if the aggregate instance is modified via one iterator, the use of the other iterators may produce undefined results afterwards.

Further SDAI operations are presented in Section 6.

5 An Example Session

Each SDAI operation belongs to a session. A *session* is the set of operations and implementation-specific activities that occur when one application uses an SDAI implementation for an unbroken period of time. Each session is initiated by an OpenSession operation and terminated by a CloseSession operation.

Figure 4 shows the first part of an example SDAI session. First the session is initiated, repository `R1` is opened, and sdai model `M1` in this repository is opened. `M1` is part of a schema instance that has the schema `example` in Figure 1 as underlying schema. Subsequently, the entity instances listed in Figure 2 are entered in the sdai model `M1`. After that, some operations are performed to find out the name of a port in the port_group. Since the attribute `members` of `port_group` is an aggregate (`SET OF port`), an iterator is used for this purpose.

6 New Features

The definition of the SDAI is still under development. In 1993 an ISO Committee Draft was sent out for ballot. After the ballot, a number of new features were added to the SDAI in order to meet the user requirements. The new SDAI version provides a basic transaction mechanism, supports queries, and enables data sharing.

Initiate session, open repository R1 *and sdai model* M1
```
session <- OpenSession()
OpenRepository(session,R1)
StartReadWriteAccessMode(M1)
```
Enter the entity instances listed in Figure 2
```
p1 <- CreateEntityInstance(port,M1)
PutAttribute(p1,name,"port1")
p2 <- CreateEntityInstance(port,M1)
PutAttribute(p2,name,"port2")
pg1 <- CreateEntityInstance(port_group,M1)
m1 <- CreateAggregateInstance(pg1,members)
Add(m1,p1)
Add(m1,p2)
```
Find out the name of a port in the port_group pg1
```
m1 <- GetAttribute(pg1,members)
i1 <- CreateIterator(m1)
Next(i1)
p1 <- GetCurrentMember(i1)
```
GetAttribute(p1,name) *returns the name of the port*

Figure 4: Sample SDAI session

6.1 Transactions

The SDAI transaction concept is designed to support an application programmer performing complex data transformations on highly interconnected data. As such, an SDAI *transaction* is defined as a sequence of operations working on entity instances. Which entity instances are available for inspection or modification is defined by the session in which the transaction is started.

The SDAI provides four transaction operations: StartTransaction, Abort, Commit, and EndTransaction. The StartTransaction operation initiates a transaction either in read-write mode or in read-only mode. After starting the transaction, entity instances might be accessed or modified. The Abort operation discards all changes which happened since the last StartTransaction operation or Commit operation. The Commit operation makes all changes persistent which happened since the last StartTransaction operation, Abort operation, or Commit operation. Neither the Commit operation nor the Abort operation terminates a transaction. Only the EndTransaction operation ends the sequence of operations started with the StartTransaction operation. The EndTransaction operation comes in two flavors: either EndTransaction with commit or EndTransaction with abort.

If an SDAI implementation supports transactions, all the operations on entity instances can only be performed within a transaction. Figure 5 shows the sample SDAI session of Figure 4 with transaction operations. In fact, the SDAI specification defines three levels of SDAI implementations: no transactions supported (level 1), transactions supported at sdai-model level (level 2), and full transaction support (level 3). In the case of

```
Initiate session, open repository R1 and sdai model M1
session <- OpenSession()
OpenRepository(session,R1)
StartReadWriteAccessMode(M1)
Enter the entity instances listed in Figure 2
t_id <- StartTransactionReadWriteAccessMode(session)
...
EndTransactionAccessAndCommit(t_id)
Find out the name of a port in the port_group pg1
t_id <- StartTransactionReadOnlyAccessMode(session)
...
```

Figure 5: Sample SDAI session with transaction operations

level 3, the operations StartTransaction, Abort, Commit, and EndTransaction are provided. In the case of level 2, the operations SaveChanges and UndoChanges are provided. In this case, `StartReadWriteAccessMode(M1)` starts a transaction on the sdai model `M1`, and the operations `UndoChanges(M1)` and `SaveChanges(M1)` can be used to abort, and commit, respectively. The transaction on the sdai model `M1` is ended automatically by `EndReadWriteAccessMode(M1)`.

6.2 Query Support

The query operation of the SDAI queries a set of entity instances and returns those entity instances meeting the specified criteria. For example, `SdaiQuery(s,"('b' <= port.name) AND (port.name < 'c')",l)` queries the set `s` and adds all the instances of `port` whose name starts with 'b' to the list `l`.

All the members of the set `s` must be instances of the entity `port` or one of the subtypes of this entity. For example, `s` might be the set of all the instances of `port` in some sdai model. For each sdai model and each entity of the underlying schema, the SDAI provides a variable, called *folder*, that represents the set of all the instances of the entity in the sdai model.

The operation SdaiQuery does not create a list of the entity instances meeting the criteria, but adds those entity instances to an existing non-persistent list. This list can be used for further SdaiQuery operations (either as set that is queried or as list to which the entity instances meeting the criteria are added) and other SDAI operations.

6.3 Data Sharing

Furthermore, the SDAI now allows that schema instances share data. I.e., an sdai model can be in more than one schema instance. Figure 6 illustrates a situation where schema instances share an sdai model.

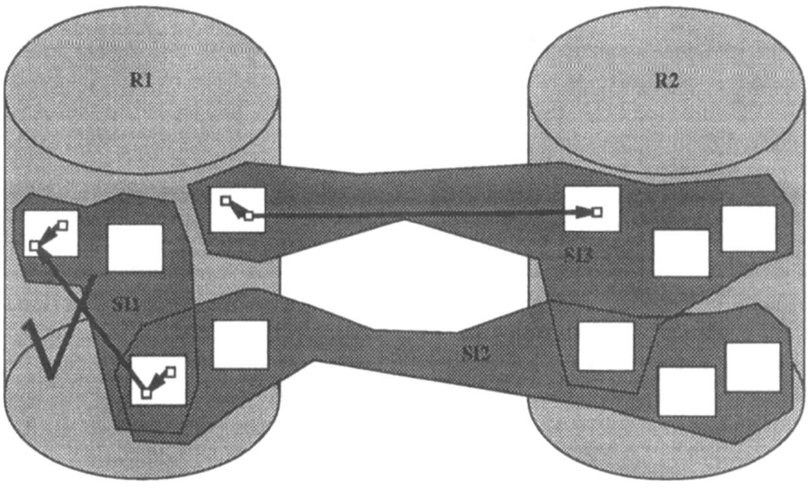

Figure 6: Data sharing

Schema instances can also share an sdai model if the schema instances have different underlying schemas. For example, suppose schema instance SI1 has AP210 as underlying schema and schema instance SI2 has AP214 as underlying schema. If the product described in SI1 and the product described in SI2 have the same product structure, they can share the sdai model describing the product structure.

Data sharing between schema instances with different underlying schemas is only possible if the shared data conform to a common subset of the underlying schemas. Such common subsets are defined by data sharing tables. For example, a data sharing table may define that the subschema "Product structure configuration", which is used by both AP210 and AP214, is a common subset of AP210 and AP214.

Note, that data sharing may introduce references from one schema instance into another. The reference indicated in Figure 6 is legal, since it is a reference inside schema instance SI1. However, at the same time, it is a reference from schema instance SI2 into SI1. In this situation any operation that is applied to schema instance SI2, like a rule validation, treats this reference as a NULL reference.

7 Conclusion

This paper presented the Standard Data Access Interface SDAI. The SDAI provides a standardized database interface to all kinds of CAx data. We believe that SDAI will also be the leading standard interface for ECAD databases.

The definition of the SDAI is now on its way to become an ISO standard. It will be a Draft International Standard (ISO/DIS 10303-22) in the first half of 1995. For achieving this status, several features had to be added or improved to the earlier SDAI version (the ISO Committee Draft sent out for ballot in 1993). A proper distinction between the physical data organization (repositories) and the logical data organization (schema instances) was introduced and data sharing was enabled, a basic transaction mechanism was added, the query operation was added, and many other improvements were made. Nevertheless, some aspects are still under discussion.

Most people involved in STEP agree that data sharing between applications using different APs should be possible. Therefore, the SDAI enables data sharing. Common subsets of APs are defined by data sharing tables (cf. Section 6.3). However, within the STEP community, there is no common agreement on how the common subsets of different APs should be described. The *Application Interpreted Construct (AIC)* was introduced for this purpose, but currently the AICs are still under development. Consequently, the actual SDAI version does not describe how the data sharing tables are populated.

Furthermore, although the meta data in the actual SDAI version provide all the necessary information, there is an ongoing discussion on which information from the EXPRESS schemas should be stored how.

References

[1] R. Anderl, "STEP - Grundlagen der Produktmodelltechnologie", in: W. Stucky, A. Oberweis (eds.), in Proc. der *Datenbanksysteme in Büro, Technik und Wissenschaft*, Springer, 1993.

[2] ISO TC184/SC4, *Product Data Representation and Exchange - Part 1: Overview and Fundamental Principles*, ISO/DIS 10303-1, ISO, 1993.

[3] ISO TC184/SC4/WG5, *Product Data Representation and Exchange - Part 11: The EXPRESS Language Reference Manual*, ISO/DIS 10303-11, ISO, 1993.

[4] ISO TC184/SC4/WG7, *Product Data Representation and Exchange - Part 21: Clear Text Encoding of the Exchange Structure*, ISO/DIS 10303-21, ISO, 1993.

[5] ISO TC184/SC4/WG7, *Product Data Representation and Exchange - Part 22: Standard Data Access Interface Specification*, ISO TC184/SC4/WG7 N350, ISO TC184/SC4/WG7, 1993.

[6] J. Owen, "STEP - An Introduction", *Information Geometers*, Winchester, 1993.

PART NINE

Design Representation and Interfaces

23

Towards Design Formalization to Support Reuse in ECAD Frameworks

Joachim Altmeyer, Bernd Schuermann, Martin Schuetze
University of Kaiserslautern, Germany

Abstract

'Design Reuse' is one of the hopeful techniques to increase the productivity in ECAD frameworks. To support this reuse we propose a formal model to describe design objects, design processes, and requirement specifications. Using this model we introduce a measure to determine reuse candidates from the design database. We address the consideration of special CAD properties and discuss appearing problems.

1. Introduction

The reuse of designs or parts of designs seems to be "*the key enabler for system design in the 90's*" [DAC94]. This reuse should comprehend all design informations: design objects as well as specifications, documentations, activities, and whole design processes. To turn this effort into reality we need an adequate formalization of the whole design. Even if our goal must be to support the reuse of all design data, we will mainly focus on the reuse of design objects and design specifications in this paper.

In section 2, we gather the requirements of a comprehensive reuse in ECAD frameworks. Section 3 focuses on the formalization of designs to support the retrieval of suitable reuse candidates. We address important problems and we investigate which of these problems can be solved by our model. In section 4, we discuss the application of this model and we present a survey of interesting current and future works.

2. Reuse in ECAD

To date, reuse is only supported rudimentarily in existing ECAD frameworks. It is restricted to the utilization of standards (e.g. the IEEE floating-point standard) or building generic descriptions of design objects or processes (e.g. generic nets in [BHK87]) [GiC93].

Let us collect properties which distinguish reuse in CAD, especially in ECAD, from many other reuse situations:

- Design Space

  Design objects in the design database are part of dependency networks (see figure 1): First, design objects are often aggregates of other design objects (*part-of hierarchy*). Second, every object has a history which expresses its evolution (*design hierarchy*). Third, there are similarities of independently designed objects defined by their properties (*property* or *feature network*). For instance, some layouts have the same technology or they concern to the same library of arithmetical functions fulfilling the IEEE floating point standard.

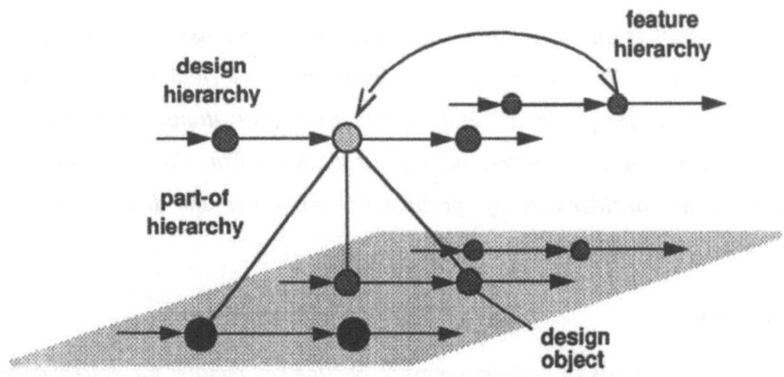

Figure 1: Design Space

- Complex Objects

  Design objects are complex objects in the sense that they own non atomic attributes, e.g. a netlist or a whole layout description.

- Requirement Specification

  Every design is initiated by a requirement specification which must be described in an adequate form.

- Design Process

  Design processes are complex processes (e.g. the processes are nested).

- Design Environment

  To describe a design in its entirety a lot of related documents is necessary (e.g. a document that describes the role of the involved tools or designers [CFI91], [HRS92]). In addition to this, there is a lot of knowledge which is not available by computers (e.g. the designers knowledge).

- Amount of Data

  The amount of data in ECAD is immense.

We can distinguish different reuse methods. Reuse can be realized *by instantiation*, *by parameterization* (*generalization*), or *by adaptation* [GiC93] [AOS94]. At reuse by instantiation, the reuse of optimized and validated objects is supported by a standardization of these. Then, these objects become types and can be instantiated within new designs. One precondition is that the underlying data model supports instantiation. For this, many suitable data models were proposed at earlier ECAD conferences. At reuse by instantiation, the creative part is to find suitable components and to integrate them well into the current design.

Furthermore, it is possible to describe components generically. For instance, we can describe an adder with the bit-width as a formal parameter. We call this reuse by generalization or parameterization. Here, the creative part is to generalize the description of a component by creating formal parameters.

In the last case, already designed solutions are reused by adapting these to the current design situation. At this reuse by adaptation, it is the goal to retrieve suitable components and to support the designer for adapting these components. To find suitable components means that the found solutions must fit the given requirement specification best. In contrast to the reuse by instantiation, the found components are then fitted to their environment not vice versa. This reuse process is outlined in figure 2. In this paper, we focus on the description of design objects, requirement specifications and design processes to support the identification of suitable reuse candidates.

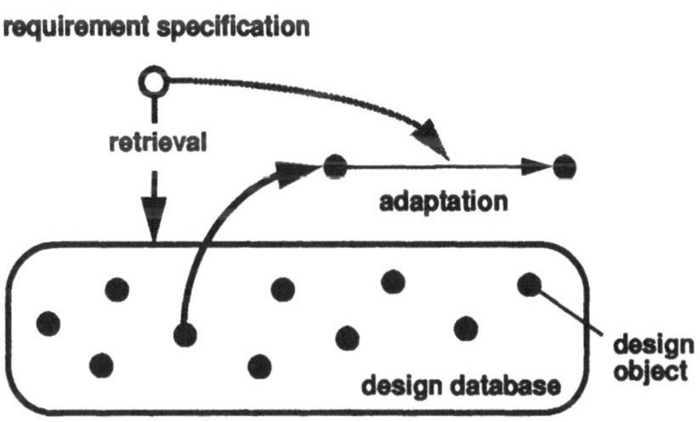

Figure 2: Reuse by Adaptation

3. Reuse Model

To support reuse in ECAD regarding the properties of the previous section we need an adequate formalization of the design, especially of the design objects, specifications, and design processes. First design formalizations without considering reuse can be found in [Yos81], [Ger90],

or [Sie91]. In this section, we present the formalization of our feature-based reuse model in detail. In a first step, we focus on the description of design objects, requirement specifications, and processes.
To describe design objects and specifications we propose two definitions:

Definition 1 (feature or property)
x is a design object or a design specification. p is a *feature* (*property*) of x iff p is a predicate with p (x) is true. ■

Definition 2 (feature set)
The *feature set* of a design object or a design specification x is defined as $P_x = \{p \mid p(x)\}$. ■

These features do not depend on the data representation of the object. Our feature notion is a generalization of the notion 'feature' or 'form-feature' in geometric CAD. The set of all possible features is divided into some clusters which we call *feature classes*. This helps us to organize features concerning the same domain. Examples of feature classes are function, bit-width, arity, technology, and area.
We extended the definition 1 and 2 from design objects to requirement specifications because even if there is a difference between specifications and design objects in their role within the design process, there is no difference in their characterization using feature sets.
In every feature class, a generalization-specialization relation '<' is defined. For instance, a feature 'Arity is 4' is a specialization of a feature 'Arity is Even', i.e. 'Arity is 4' < 'Arity is Even'. Using this relation the instances of each feature class builds a mathematical lattice structure with the least element 'no specification' and a greatest element which spans the whole definition space of the feature class [OnY87].
Now, we define generic features:

Definition 3 (generic features)
A *generic feature* g defines a set G of features with $G = \{p \mid p < g\}$ where p is a feature. ■

Using this definition we are able to handle a generic feature g within a feature set P_i in the way that we replace it by the features of the corresponding set G which results in a more simple and uniform model:

$$P_i' = (P_i - \{g\}) \cup G$$

Before we define similarity functions between design objects and specifications, we describe the correlations between the definitions above and the design process:

Definition 4 (successful design)
s is a design specification. The design is *successful* iff there exists a design object x with $P_s \subseteq P_x$. ■

Definition 5 (refinement design step)
x and y are design objects and s is a design specification. φ is a *refinement design step* iff $\varphi(x) = y$ and $(P_s \cap P_x) \subseteq (P_s \cap P_y)$. ■

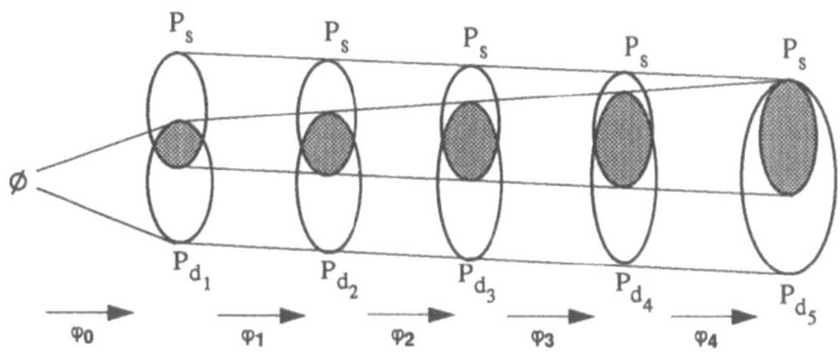

Figure 3: Venn Diagrams of Design Refinement Steps
P_s is the feature set of the overall requirement specification s, φ_i are refinement steps, P_{d_i} are the feature sets of the design objects d_i

Or with other words: "With refinement steps, we come closer to fulfil the overall design specification". Figure 3 illustrates the evolution process. These refinement steps define the design hierarchy which describes the evolution of the design. Every object, which results from a refinement design step, can be reused by itself. This is the reason why we call every object resulting from a refinement step a design object. It is important that, during the design, there are other steps, e.g *correction steps* or *analyzing steps*.

Example 1 (design hierarchy)
In figure 4, the design hierarchy is illustrated by an example using the PLAYOUT design system [Zim89]. In a first step φ_0, a VHDL program d_1 was written from scratch ($d_0 = \emptyset$). In step φ_1, a high level synthesis tool was applied resulting with a netlist d_2 as a result. Then, two alternative floorplans d_3 and d_4 were designed. Using these, two layouts d_5 and d_6 were assembled. The overall design was successful regarding the requirement specification s iff $P_s \subseteq P_{d_i}$ for a design object $d_i \in \{d_1, d_2, \dots d_6\}$. ■

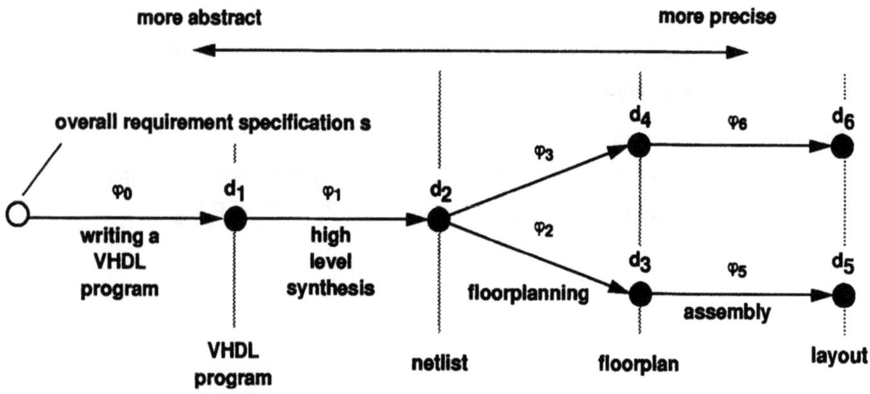

Figure 4: Example of a Design Hierarchy

So far, we introduced a characterization of design objects and specifications. In the following, we focus on their similarities. To determine the similarity of features we use individual (asymmetric) similarity functions for each feature class:

Definition 6 (feature similarity function)
C is a feature class. The *feature similarity function* sim_C is defined as $sim_C\colon C \times C \rightarrow [0, 1]$. If $sim_C(p, q) = 1$, then the feature p is equivalent to the feature q. In the case of $sim_C(p, q) = 0$, p is totally different from q. ■

The feature similarity function $sim_C(p, q)$ roughly expresses the expense of converting the feature q to the feature p. We define a similarity function SIM_C for a feature class:

Definition 7 (feature similarity function of a feature class)
If the feature sets $P_s^C = \{p \mid p \in P_s \wedge p \in C\}$ and $P_x^C = \{p \mid p \in P_x \wedge p \in C\}$ contain only one element, the *feature similarity function of the feature class* C, SIM_C, is sim_C with the instances of P_s^C and P_x^C, respectively. Otherwise, we use the following function:

$$SIM_C(s, x) = \frac{\sum_{p_s \in P_s^C} \max\left(\bigcup_{p_x \in P_x^C} \{sim_C(p_s, p_x)\} \right)}{\left|P_s^C\right|}$$ ■

Now, we define a similarity function SIM for design objects which enables us to determine the reusability of design objects regarding a given specification:

Definition 8 (reusability function)

s is a design specification, x a design object, and Γ the set of all feature classes with instances in P_s. The *reusability function* SIM is defined as:

$$SIM(s, x) = \sum_{C \in \Gamma} r_C \cdot SIM_C(s, x).$$

$r_C \in \mathbb{R}^+$ are *relevance factors* (weight factors) which express the importance of the corresponding feature classes for the comparison (the sum of all weight factors has to be 1). ∎

Using the extended function SIMC of definition 7, we are able to consider generic design objects, i.e. objects with generic features, as well as multi-functional design objects, i.e. objects with more than one feature at the feature class function.

Example 2 (multi-functional design object)

If we search for a multi-functional design object with specification s and P_s = {'Function is Add', 'Function is Multiply', 'Width is 32', 'Arity is 2'} and we find a design object x with the feature set {'Function is Add', 'Width is 32', 'Arity is 2'} then $SIM_{Function}$ (s, x) = 1/2 if $sim_{Function}$ ('Function is Multiply', 'Function is Add') = 0. ∎

Regarding the definition of P_s^C and P_x^C, resp., (see above) we define the context P_i^C of a feature set as:

Definition 9 (context of a feature set)

$\bar{P}_i^C = P_i - P_i^C$ is the *context* of P_i^C. ∎

One of the problems with the formula of definition 8 is that each feature set P_s^C and P_x^C, resp., is investigated independently of its context. To consider the context of features we do the following:

- ❑ Selected metrics consider more than one feature class at a time. For instance, the feature classes x-dimension and y-dimension, i.e. the feature classes area and aspect ratio are considered together. Figure 5 shows an example of considering these classes.
- ❑ To express facts like "A n-input multiplexer can be constructed of two n/2 multiplexer and one 2-input multiplexer" we propose substitution rules which cause new search processes in addition to the current search. Substitution rules are represented as follows:

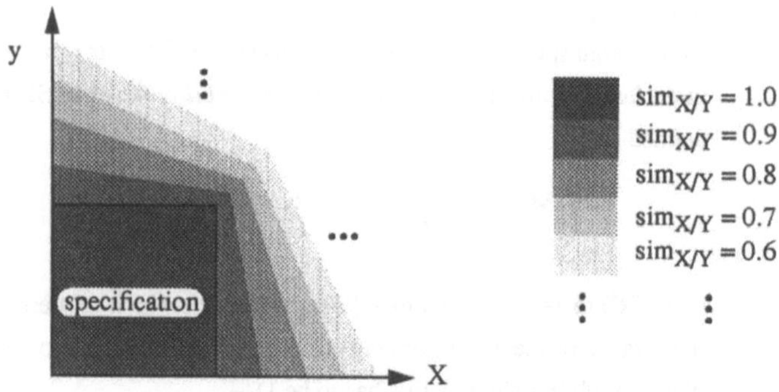

Figure 5: Considering the Two Feature Classes X-Dimension and Y-Dimension at a Given Specification

$$P \rightarrow P_1 \;\&\; P_2 \;\&\; \ldots \;\&\; P_n.$$

If we search for a design object using a specification s with the feature set P_s, the rule fires if the LHS P matches P_s with $P \subseteq P_s$. Then, n new search processes are started with new specifications s_i, i = 1, ..., n, with:

$$P_{s_i} = (P_s - \{P\}) \cup P_i$$

The overall retrieval yield a result if either the old search process finds a suitable object or all new search processes caused by the RHS yield a result. The following example motivates this strategy:

Example 3 (substitution rule)

If we search for a multiplier with the specification {'Function is Multiply', 'Width is 16', 'Arity is 2', 'Technology CMOS'} and we apply the substitution rule (see also figure 6):

{'Function is Multiply', 'Width is n', 'Arity is 2'} →
{'Function is Add', 'Width is n', 'Arity is 2'} &
{'Function is Store', 'Width is n', 'Arity is 1'} &
{'Function is Shift-Right', 'Function is Store',
'Width is 2n', 'Arity is 1'};

then three new search processes are started with the specifications s_1 = {'Function is Add', 'Width is 16', 'Arity is 2', 'Technology CMOS'}, s_2 = {'Function is Store', 'Width is 16', 'Arity is 2', 'Technology CMOS'} and s_3 = {'Function is Shift-Right', 'Function is Store', 'Width is 32', 'Arity is 2', 'Technology CMOS'}. ■

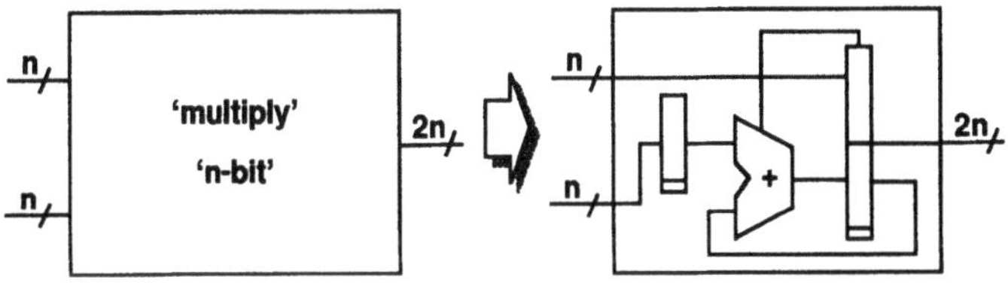

Figure 6: Substitution of a Multiplier by Network of other Modules

With these substitution rules, we exploit the part-of hierarchy. Another problem is the consideration of the design history. Using special metrics on the design state we exploit that adaptations can easier be made in more abstract descriptions than in more specific descriptions of design objects (see figure 7).

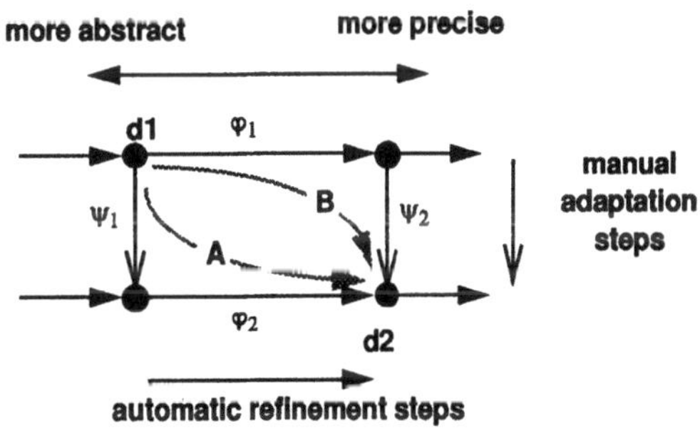

Figure 7: Use of the Design Hierarchy
It is obvious that, in this case, the way A is easier to realize than the way B.

4. Discussion

The presented formalization is implemented in our prototype RODEO [AOS94]. (RODEO is an acronym for reuse of design objects.) As a part of the VLSI design system PLAYOUT [Zim89], RODEO supports designers in finding similar design objects and specifications. RODEO interactively delivers a set of reuse candidates which fit the current requirement specification best. RODEO has only an abstract view of the design data in accordance with the definitions 1 and 2, i.e it does not consider detail data as netlists or layouts but only abstract features. It is necessary that the designer investigates the retrieved candidates in detail afterwards. Therefore, RODEO helps the designer and does not replace him.

Although the quality of our approach is attested by our implementation, our model does not consider all properties of section 2. Therefore, we will finally gather the problems which must be solved toward a comprehensive reuse in ECAD: Firstly, we should minimize the checks by the designer after the retrieval process. Secondly, even if we use substitution rules the consideration of the feature context is not sufficient. It must be the goal to find context-sensitive descriptions of similarities. Thirdly, suitable domain-dependent metrics for the similarity function sim_C for one or more feature classes must be determined. To solve this problem we must study many experiments with design reuse. Fourthly, even if RODEO contains an explantation component to support the designer using the retrieved design objects, more information should be gained towards an automatic adaptation step. And finally, the reuse of other design informations must be supported.

References

[AOS94] J. Altmeyer, S. Ohnsorge, B. Schürmann, "Reuse of Design Objects in CAD Frameworks", Proc. Int. Conference of Computer Aided Design, 1994

[BHK87] B. Becker, G. Hotz, R. Kolla, P. Molitor, "Hierarchical Design based on a Calculus of Nets", Proc. 24th Design Automation Conference, 1987

[CFI91] "Tool Encapsulation Specification", CFI Document No. 51, 1991

[DAC94] "Design Reuse: Fact or Fiction?", Panel, Proc. 31th Design Automation Conference, 1994

[GiC93] E. Girczyc, S. Carlson, "Increasing Design Quality and Engineering Productivity through Design Reuse", Proc. 30th Design Automation Conference, 1993

[Ger90] J. Gero, "Design Prototypes: A Knowledge Representation Schema for Design", AI Magazine, Winter, 1990

[HRS92] C. Huebel, D. Ruland, E. Siepmann, "On Modeling Integrated Design Environments", Proc. Int. European Design Automation Conference, 1992

[OnY87] M. Onosato, H. Yoshikawa, "A Framework on Formalization of Design Object for Intelligent CAD", Proc. IFIP WG 5.2 Workshop on Intelligent CAD, Boston, 1987

[Sie91] E. Siepmann, "Entwurfstheorie und Entwurfsdatenmodellierung fuer CAD-Frameworks", Ph.D. Dissertation, University of Kaiserslautern, 1991, (in German)

[Yos81] H. Yoshikawa, "General Design Theory and a CAD System", in "Man-Machine Communication in CAD/CAM", T. Sata, E. Warman (eds.), Proc. IFIP WG 5.2/5.3 Working Conference (Tokyo 1980), 1981

[Zim89] G. Zimmermann, "PLAYOUT - A Hierarchical Design System", Information Processing 89, G.X. Ritter (ed.), Elsevier Science Publishers B.V. (North Holland), IFIP, 1989

24

A Generator for Graph-Based Design Representations*

Ansgar Bredenfeld

GMD-SET, Schloß Birlinghoven, D-53754 Sankt Augustin, Germany, ansgar@borneo.gmd.de

Abstract

Tool integration services are an important means to make design tools inter-operate in a design environment. Today, the problem of fine-grained data integration is tackle from two directions. From the application domain using fixed design tool interfaces and internal forms based on specific design representations and from the database domain applying models and interfaces mainly based on semantic and/or object-oriented data models.

In our approach, we use concepts of both directions in order to further close the still existing semantic gap between a design tool and a framework database. We extend an object model comparable to existing ones by an abstract graph model which reflects structures and functionality occurring in various graph-based design representations. This extended model is used as basis for a new tool integration service. It provides a schema-driven code generator for implementations of graph-based design representations. In addition, presentation integration is achieved by providing a configurable graphical interface which directly offers visualization functionality and interactive access to fine-grained design data.

1. Introduction

Design environments consists of several design tools and services for data management and design methodology management [1]. Enabling and supporting interoperability of design tools in environments is the major goal of these framework services. Tool integration services have to offer mechanisms to easily plug tools into the environment. Tool integration may be classified in *data integration*, *presentation integration*, *control integration*, and *process integration* [2]. This paper is concerned with data integration but also considers aspects of presentation integration.

Data integration services provide tool interfaces between a common data repository of a framework and design tools. Data representations are needed as conceptual models of the data which is transferred via the interface and handled by the tools. The granularity of data representations may be used to further classify data integration in a framework.

---

*. This research was supported in part by the commission of EC under ESPRIT contract 7364

Design data on a *coarse-grained* level are used as atomic entities in the context of design flow management. Coarse-grained design representations are often associated with textual design files expressed in specific description languages e.g. VHDL [3], EDIF [4], or Verilog [5]. Tool integration on this level is accomplished by tool encapsulation techniques utilizing so-called tool wrappers. They may be regarded as a more or less complex interface structure between an encapsulated design tool and the data repository of the framework. Design data transfer via this interface is usually controlled by a design flow manager. Often tool wrappers use temporary workspaces containing design files checked out from the data repository. Utilization of the fine-grained contents of design files is only possible after parsing done by the encapsulated tool.

Design tool algorithms operate on design data of *fine granularity*. Therefore, fine-grained design representations are suitable conceptual models for the data structures of design tools as well as for the tool interfaces providing design tools with design data from a database. In contrast to that, encapsulated design tools are only provided with design files. Their coarse granularity does not reflect fine-grained dependencies within and between design files. Recent research in the area of tool encapsulation [6] has proposed to split textual design files into text chunks in order to operate on design data of finer granularity. Nevertheless, real fine-grained tool integration remains the ultimate goal to allow for tight interoperability between design tools. Fine-grained design representations are a crucial prerequisite to allow for e.g. back-annotation and highlighting mechanisms within shared fine-grained design data. Interoperability on this level of granularity is not only needed within the scope of a single design representation e.g. a netlist, but also between several design representations. As only one example, imagine a system for interactive high-level synthesis e.g. [7] or [8] which needs back-annotation mechanisms between the original behavioural specification and a design structure synthesized from it in order to allow for interactive exploration and optimization of a synthesized design. Such systems need fine-grained links between the behavioural representation e.g. a control/data flow graph and the structural representation e.g. a netlist. In future, interoperability even between hardware and software representations will be an important issue of tool integration in hardware/software codesign systems [9].

An analysis of design representations shows that they are often based on more or less complex graph structures. This motivates our approach to identify some of their common structures and functionality needed by design tools for their manipulation. Since several data models for fine-grained tool integration exist [10,11,12,13,14], we use them as starting point for our model. We extend their modelling concepts with graph structures and functionality in order to offer a powerful tool integration service. This approach will further close the still existing semantic gap between database models and the often graph-based conceptual model of design tools. In addition, reuse of graph functionality will reduce the code needed to be written for integrating existing tools and especially for constructing new tools.

This paper is structured as follows. Section 2 gives an overview of related work. Section 3 briefly describes our abstract graph model. Section 4 focuses on a new tool integration service which allows to automatically generate implementations of and graphical interfaces for graph-based design representations. Section 5 gives a first case study on using a generated netlist representation to integrate a multi-level fault simulator, an existing Verilog parser, and a graphical interface which is provided by our tool integration service. Section 6 concludes the paper with an outlook to future work.

2. Related Work

Standardized fine-grained design representations as basis for tool integration exist for netlists only. The CAD Framework Initiative, Inc. (CFI) has defined the Design Representation Programming Interface (DRPI) [15] which is a strongly typed interface between an integrated design tool and a framework database. Netlists are an important design representation based on graphs. Nevertheless, graph-based design representations also occur in other tool domains. Control flow graphs, data flow graphs, or state transition diagrams are only a few examples for them found in the high-level synthesis domain. They serve as conceptual models of internal forms in specific high-level synthesis systems e.g. [7] or [8].

Several data models for fine-grained tool integration have been proposed in recent years [10,11,12,13,14]. They are mainly based on object-oriented or semantic data models. The structure of design data is described in a schema. Design data stored in a framework database have to conform to this schema. Manipulation operations are provided to handle these design data. The models described in [13] and [14] use a small set of generic operations. Object types are passed as parameters to their interfaces. Therefore, type checking has to be done dynamically which is critical with respect to tool performance.

Although these data models made a first step to narrow the semantic gap between a simple internal model of a database and the complex conceptual model of a design tool, all of them lack explicit support of graph structures. Nevertheless, many conceptual models of design tools are based on more or less complex graph structures as shown before. In order to further close the still existing semantic gap between tools and databases, we extend an object model comparable to these existing models by an abstract graph model to allow for homogeneous modelling of complex graph-based design representations.

3. Abstract Graph Model

The object model we use as basis supports modelling concepts known from object-oriented programming languages like object types (classes), attributes (data members), inheritance (class hierarchy), polymorphism and additionally object identity known from object-oriented databases (OODB). Furthermore, we add concepts from semantic data models by supporting different types of relationships between object types (e.g. unidirectional and bidirectional references as well as containment relationships; each of them alternatively with one-to-one, one-to-many, or many-to-many cardinality). Language constructs representing these modelling concepts are used to express design data in a schema.

Operations allow for creation, deletion, navigation, and manipulation of design data conforming to the schema. We abstract these operations by a small set of formally defined generic functional primitives. They are represented in an intermediate form to be independent of a specific implementation base. This approach allows us to map functional primitives to either an object manipulation language of an OODB or to efficient method implementations operating on C++ in-memory data structures.

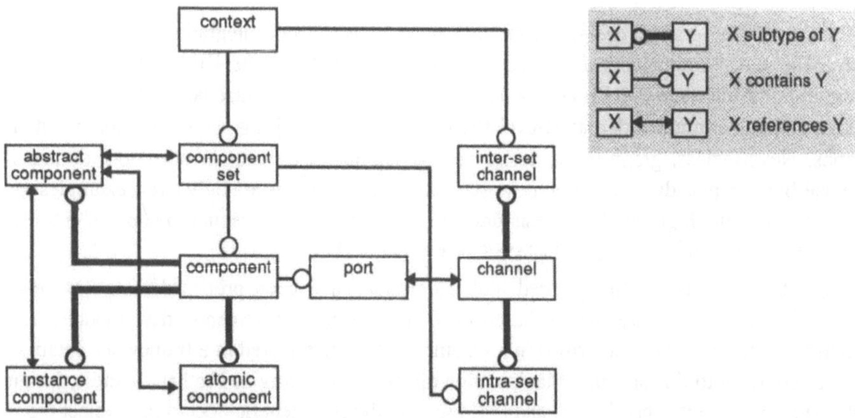

Figure 1. Abstract graph schema

The language constructs of the object model are used to define the structure of our abstract graph model. It is given by the simplified *abstract graph schema* depicted in figure 1. A sample graph structure conforming to the abstract graph schema is illustrated in figure 2.

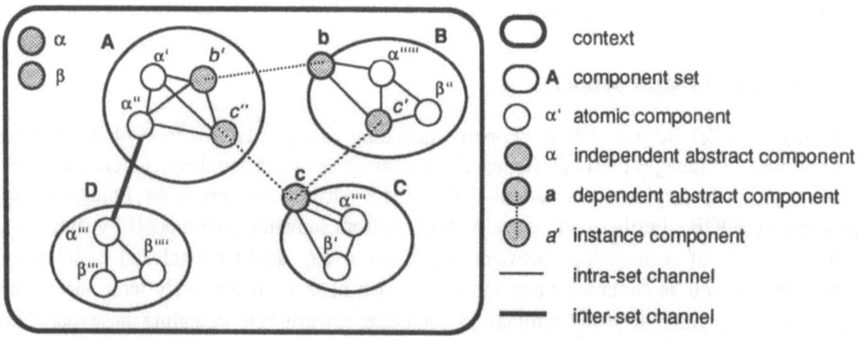

Figure 2. Sample graph structure

The central entity is a *component* having an interface structure consisting of a set of directed *ports*. Connectivity between components is modelled by a reference from each port to a *channel.* Channels may reference many ports. They are the communication links between components. Signals may be transported via channels between components.

Abstraction is modelled by an *abstract component.* In general, abstract components provide descriptions of the interface structure and the behaviour of components derived from them. Two different abstraction mechanisms are supported. *Independent abstract components* (α, β) may be regarded as type descriptions of *atomic components* (α', α",..., β', β",...). Therefore, atomic components derived from the same independent abstract component have the same interface structure (α has 3 ports, β has 2 ports). These few object types are sufficient to model flat graphs of typed multi-port components connected by multi-channels. They are suitable to model e.g. flat netlists or flow graphs.

Additional concepts needed by graph-based design representations extend this basic graph structure. Grouping of components in a flat graph is done by disjoint *component sets* (*A*, *B*, *C*, *D*). They partition a flat graph. As a consequence channels are specialized into two disjoint sets of *inter-set channels* and *intra-set channels* (cf. figure 2).

The second abstraction mechanism is used to construct hierarchies. It allows to abstract a group of components (component sets) by a single *dependent abstract component* (*a*, *b*). This kind of abstract component is not an isolated type description as an independent abstract component but rather represents the interface structure of the abstracted component set. Structural hierarchies are constructed by deriving several *instance components* (*b'*, *c'*, *c''*) from dependent abstract components. Instance components have the same interface structure as their describing dependent abstract components.

Components are clustered in component sets which are in turn clustered on a higher level by *contexts*. This supports semantic clustering of graphs already in the model and is exploited to map graphs and their subgraphs to clusters in an underlying database.

Graph functionality is specified using the abstract graph schema as structural basis. Functionality includes e.g. definition of independent abstract components, graph construction and destruction by creating and deleting atomic components, connecting and disconnecting components, and arbitrary navigation in component graphs. Creation of dependent abstract components from component sets in conjunction with subsequent derivation of instance components is used to dynamically construct structural hierarchies. All transformation operations preserve integrity of connectivity between components.

In addition, functionality to support construction of design tools is provided, e. g. hierarchy transformations (folding, unfolding), signal flow analysis (levelizing), signal propagation, and behavioural descriptions of abstract components. A detailed description of this functionality is beyond the scope of this paper.

4. Model-Based Tool Integration Service

The abstract graph model was designed to improve fine-grained tool integration and to speed up tool construction. Therefore, we have to provide a process to make the abstract model usable as a tool integration service offering graph structures and functionality to a tool integrator or a tool constructor.

4.1 Design Representation Schema

More complex design representations consist of several interrelated graph-based structures. For example control/data flow graphs (CDFG) are used as behavioural representation of a design in the high-level synthesis domain. These CDFGs are a combination of two graphs - a control flow graph and a data flow graph. In order to be capable of modelling such structures we allow to use the abstract graph schema several times in a single schema. This mechanism may be considered an instantiation of the abstract graph schema on schema level. Therefore, we call occurrences of the abstract graph schema *graph schema instances.* Since each specific graph structure has its own interpretation of the abstract graph schema, we bind each object type of the abstract schema to a specific object type in a graph schema instances. Thus, each graph schema instance provides a specific graph structure with a set of well defined graph functionality.
Graph schema instances may be arbitrarily extended using the language constructs provided by the object model. Since graph schema instances are described in terms of these language constructs, object types, relationships, and attributes are simply added to the schema. Furthermore, object types may be specialized. The resulting schema containing several graph schema instances with arbitrary extensions and specializations is called a *design representation schema.* It is used as input for the code generator.

4.2 Code Generator

We have implemented a prototype of a new tool integration service based on the abstract graph model. Its overall architecture is given in figure 3.

The code generator is capable of creating implementations of design representations described by a design representation schema. It allows to map a schema to an object-oriented database and to an C++ in-memory data structure. The first is important for concurrent shared access to a design database, the latter crucial for efficient in-memory databases to gain acceptable tool performance. Since performance is regarded a key requirement especially for tool construction, we do not pass object types as parameters to a generic tool interface as done in [13] or [14]. Instead, we avoid dynamic type checking by generating a set of strongly typed operations for the object types of the schema. We use the generic functional primitives of the object model as templates for these operations. This avoids the performance degradation caused by dynamic type checking but keeps the advantages offered by a flexible schema description.

During the generation process, operations are mapped to in-memory C++ data structures with manipulation methods or to calls of the object manipulation language of an object-oriented database. We are able to easily adapt our schema to a schema of an OODB by providing syntactical mappings of our language constructs and generic functional primitives to the data description language (DDL) and the data manipulation language (DML) supported by the OODB. Since our object model supports structural elements and operations of the ODMG object model [16], this mapping is more or less straight forward.

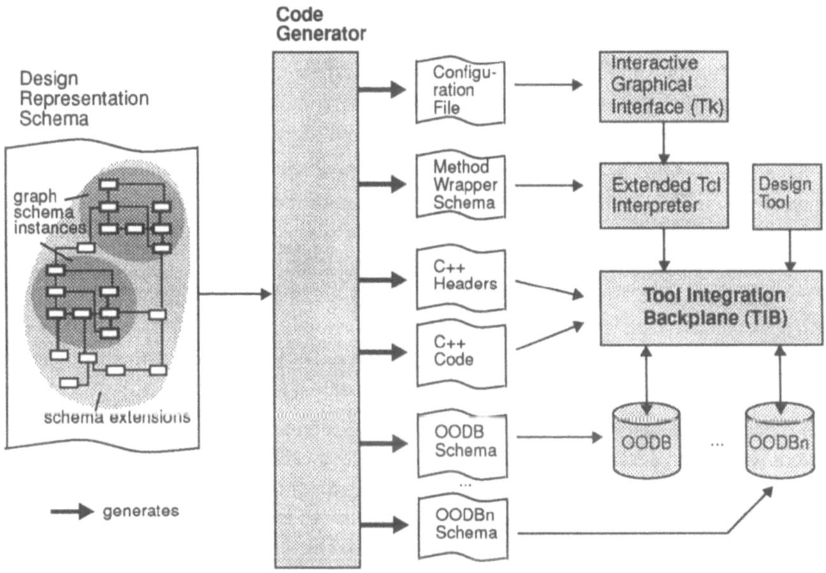

Figure 3. Overall architecture of the tool integration service

The specification of graph functionality is based on the functional primitives belonging to the abstract graph schema. This specification in intermediate form is taken as a kind of template for the methods implementations for each graph schema instance of the design representation schema. In addition, each extension to the schema results in further methods for their handling. All these methods are provided to the tool integrator or tool constructor as a generated C++ class library. We call these libraries *tool integration backplanes* (*TIB*). A design tool only needs to include a generated header file and to link the generated library.

As an additional feature of our tool integration service, the code generator produces a wrapper description for the generated methods. This *method wrapper schema* is used to bind our methods as extensions to the interpreted language Tcl [17]. Thus, all generated methods are offered as Tcl commands to the tool integrator, tool constructor, or even the end user. This offers interactive textual access to a database containing design data conforming to the design representation schema.

4.3 Graphical Interface for Design Representations

Presentation integration has the ultimate goal to provide common interaction and presentation paradigms in the whole design environment [2]. Unified graphical data presentations and homogenous interaction mechanisms are suitable means to achieve this goal.

We consider presentation integration by providing a graphical interface for data visualization as an integral part of our tool integration service. Offering such an interface is a way to separate visualization functionality from the pure algorithmic kernel of a newly constructed design tool. Existing tools which are integrated using our tool integration backplanes may also profit from means for design data visualization as shown in our case study.

We have implemented an interactive graphical interface based on the extended Tcl and the graphical toolkit Tk [17]. It exploits the structure of the abstract graph model and offers a set of various display routines to visualize graphs. This approach allows to implement visualization functionality only once for the abstract graph schema and reuse it in each graph schema instance. Since a design representation expressed by a graph schema instance may prefer a specific visualization policy, we allow to control displaying of graphs. This is accomplished by generating a configuration file for the graphical interface. The tool integrator or constructor may change its default settings to achieve a specific visualization for each graph schema instance.

Each object of a graph structure is associated with a pop-up menu which offers the set of methods defined on the corresponding object type. Selection of a menu entry directly issues the corresponding method on that object. The tool constructor or tool integrator may add entries to this pop-up menus. Actions executed after selection of added menu entries are specified by scripts in extended Tcl. This allows to arbitrarily extend the graphical interface to the needs of the tool integrator or tool constructor.

The effort for getting interactive graphical access to a specific design database is now reduced to changing and extending the generated default configuration file of the graphical interface. The configured graphical interface then serves as an interactive tool to the design database. It allows for data input, tracing of method invocations, as well as querying the database.

5. Tool Integration Case Study

In a first case study, we integrated a state of the art multi-level fault simulator [18] with an existing Verilog [5] parser. We define a design representation schema for hierarchical netlists using a single graph schema instance. The entities of the netlist are bound to object types of the graph schema instance. For example an instance is modelled either by an atomic component (instance of a leaf cell) or an instance component (instance of a composite cell). Composite cells are modelled by a component set with a dependent abstract component. The first models the aggregation aspect, the latter the abstraction aspect of a cell. By this, semantically different aspects are explicitly separated. Nets are mapped to channels.

This design representation schema is fed to the code generator which produces a tool integration backplane from it. The resulting class library is linked to the fault simulator and to the Verilog parser. The mapping of the internal data structure of the simulator to the generated netlist representation was straight forward since both are essentially hierarchical graph structures. We extend the internal data structure of the simulator to store back-annotation references for each instance in the netlist to allow for fine-grained interoperability. The simulator is now able to use the graphical interface which has been configured to serve as a visualization tool for hierarchical netlists.

The source code of the fault simulator and the Verilog parser had to be extended by calls of generated methods and some additional mapping code. The whole integration took only a week. The extension of the simulator data structure by back annotation references increases the memory requirements of the simulator by approx. 20% but allows for easy fine-grained interoperability. The runtime behavior of the simulator is not affected by this integration technique.

6. Conclusion

Our approach to fine-grained tool integration extends existing data models by an abstract graph model which defines common structures and functionality needed to handle graph-based design representations. In this, our model is a significant extension to the expressive power of existing models.

We implemented a tool integration service based on this model that automatically generates implementations of design representations from a schema description. Since we consider presentation integration as an integral part of tool integration, we provide a configurable graphical interface which serves as a flexible interactive visualization tool for generated graph-based design representations.

Currently, we are working on a second case study applying our model and service to tool construction. We are implementing a high-level synthesis algorithm using a generated design representation for high-level synthesis. We add two more graph schema instances for control and data flow graphs to the netlist schema used in our first case study. This will allow for fine-grained interoperability between the synthesis tool and the already integrated simulator.

References

[1] Barnes, T.J., D. Harrison, A.R. Newton, R. L. Spickelmier, "Electronic CAD Frameworks", *Kluwer Academic Publishers*, Boston, 1992

[2] Thomas, I., B. A. Nejmeh, "Definitions of Tool Integration for Environments", *IEEE Software*, March 1992

[3] *IEEE Standard 1076*, VHDL Language Reference Manual, 1993

[4] Electronic Industries Association. Electronic Design Interchange Format Version 3 0 0 Reference Manual, 1993, *EIA Standard EIA-618*

[5] Thomas, D.E., P. Moorby, "The Verilog Hardware Description Language", *Kluwer Academic Publishers*, 1991

[6] Schettler, O., S. Heymann, "Towards Support for Design Description Languages in EDA Frameworks", *Proc. of ICCAD*, November 1994

[7] Wu, A.C.H., T.S. Hadley, D.D. Gajski, "An Efficient Multi-View Design Model for Real-Time Interactive Synthesis", *Proc. of ICCAD*, 1992, pp. 328-331

[8] Ang, R.P., N. D. Dutt, "Equivalent Design Representations and Transformations for Interactive Scheduling", *Proc. of ICCAD*, 1992, pp. 332-335

[9] *Proc. of the Third Int. Workshop on Hardware/Software Codesign*, September 1994, Grenoble, France

[10] Sim, M.N., P.M. Dewilde, "An Object-Oriented Persistent Database Interface for CAD", *Proc. of EDAC*, 1990, pp. 363-367

[11] Van der Wolf, P., R. van Leuken, "Object Type Oriented Data Modeling for VLSI Data Management", *Proc. of the 25th Design Automation Conference*, 1988, pp. 351-356

[12] Heijenga, W., U. Jasnoch, E. Radeke, "DaDaMo - A Conceptual Data Model for Electronic Design Applications", *Proc. of EDAC*, 1992, pp. 394-398

[13] Schettler, O., A. Bredenfeld, "BEPPO: A Data Model for Design Representation", *Proc. of EuroDAC*, September 1993, pp. 378-382

[14] Filer, N., M. Brown, Z. Moosa, "Integrating CAD Tools into a Framework Environment Using a Flexible and Adaptable Procedural Interface", *Proc. of EuroDAC*, September 1994

[15] "Design Representation Programming Interface: Electrical Connectivity", Version 1.3.0-050694, *CAD Framework Initiative*, Inc., Austin, USA, 1994

[16] Cattell, R.G.G., "The object database standard : ODMG-93", *Kaufmann*, San Mateo, CA, 1994

[17] Ousterhout, J. K., "Tcl and the Tk Toolkit", *Addison-Wesley Publishing Company*, Reading, MA, 1994

[18] Meyer, W., R. Camposano, "Fast Hierarchical Multi-Level Fault Simulation of Sequential Circuits with Switch-Level Accuracy", *Proc. of 30th Design Automation Conference*, 1993, pp. 515-519

25

The Visual Interface of the STAR Framework [1]

Ronaldo S. Mello, Lia G. Golendziner, Flávio R. Wagner
Universidade Federal do Rio Grande do Sul, Instituto de Informática
Caixa Postal 15064, 91501-970 Porto Alegre RS, Brasil
e-mail {mello,lia,flavio}@inf.ufrgs.br

Abstract

Visual data management employed in EDA Frameworks helps in many design tasks, offering various facilities, specially related to the formulation of queries. In the STAR framework, a visual interface is the primary user interface to the system, allowing access to different managers and offering facilities for browsing and querying design objects. A graphical browser is available, allowing the navigation through design objects stored in the database, such that object properties can be shown and related objects can be investigated, departing from a selected one. Graphical queries can be formulated in an interactive way. Complex queries can be formulated via a textual SQL-like query language, extended with facilities for version manipulation and navigation in the inheritance hierarchy defined by the STAR data model.

1 Introduction

Data management in eletronic design automation (EDA) frameworks usually includes design object modelling, complex object manipulation, design alternatives representation and version control. Databases, in this kind of application, store objects with different types of relationships: composition, representation and versioning [1]. These relationships raise several query possibilities, useful in different stages of the design process.

From a database point of view, query formulation is usually done in a textual way. The disadvantage of a textual query language is that it does not apply to new or infrequent users considering that a good knowledge of the database schema and query language syntax is needed. This problem is being solved with the development of visual technology which provides user-friendly graphic interfaces and visual languages, allowing object visualization and manipulation in a graphical way.

Visual data management employed to EDA frameworks helps in many design tasks, offering various facilities, specially related to the formulation of queries. Database browsing, for example, allows the investigation of descendants in a representation hierarchy. Object selection can be provided by an interactive query formulation. In this case, the designer can describe query conditions by selecting object properties, conditional symbols and constant values, for example. All these facilities and other

[1] This work is partially supported by CNPq and CAPES.

ones are available in the **visual interface of the STAR framework**, under development at Universidade Federal do Rio Grande do Sul.

This paper is organized as follows. Section 2 describes the features and the data model of the STAR framework. In section 3, the visual interface is presented and its browsing and query formulation facilities are exemplified. Section 4 is dedicated to the STAR textual query language, section 5 compares the visual interface with other EDA framework interfaces and, finally, conclusions and future works are discussed in section 6.

2 The STAR Framework

STAR is an open system that supports integration of several tools aimed at different tasks, architectures and design technologies [2]. The STAR data model allows the definition of *object schemata* in a hierarchical way, where schema nodes can be generalizations of other nodes.

The root node in this inheritance hierarchy is a *Design* object. Each Design gathers an arbitrary number of *ViewGroups* and *Views*. Each *ViewGroup* may in turn gather, according to defined criteria, any number of other *ViewGroups* and *Views*. *Views* are of one of three types: *HDL Views*, for behavioral descriptions, *MHD Views*, for structural descriptions, and *Layout Views*, for physical descriptions [2, 3]

Associated to each View, there is a graph of *ViewStates*, where each *ViewState* contains the real design data, in one of the various design representations (layouts, HDL descriptions, and so on). Each *ViewState* inherits attributes from the corresponding *View* and its ascendants in the inheritance hierarchy (upwards to the *Design* node). A *ViewState* can be described as a composition of sub-objects, and each of these may reference objects that are represented by their own object schemata.

All nodes of the object schemata can be versioned, and the Version Manager automatically keeps track of the inheritance relationships among these versions. Design version management in STAR is described elsewhere [3].

Object schemata can evolve in time, and new schemata can be defined dynamically, departing from an existing schema and making changes to it [4]. Schema evolution is maintained through versions, so that version management is applied not only to the design objects themselves, but also to object schemata.

A configuration is defined for each *ViewState* as a selection of a particular version of a particular design object for each sub-object within the *ViewState*. Besides static and dynamic configurations, the STAR framework also supports open configurations (as in VHDL). The Configuration Manager [5] is a special module that allows specification of configurations either in an interactive way or through an application programming interface.

3 The Visual Interface

The visual interface is the primary user interface of the STAR framework, allowing access to different managers and offering facilities for browsing and querying design objects in a graphical or textual way

[6].

Figure 1 shows the visual interface architecture and its communication with the other STAR components. The visual query module is responsible for processing interface commands or events and returning objects and messages to be visualized. The designer interacts with the visual interface either through navigation (browsing actions) or formulating graphical queries. Both forms are converted to a textual query specification which is sent to the textual query module. In addition, textual queries can be directly formulated by designers which are familiar with its syntax. The textual query syntax is described in the section 4.

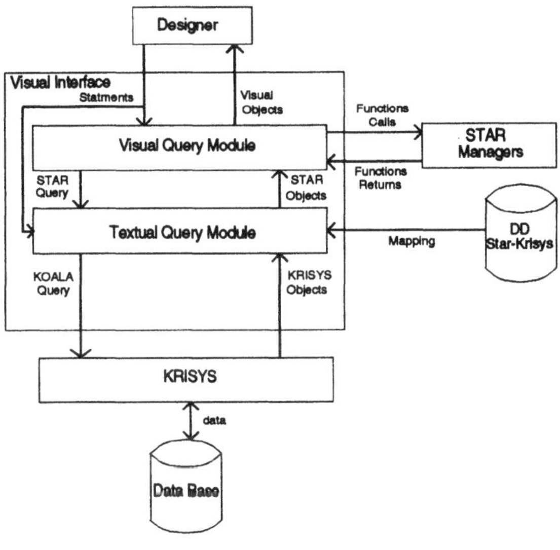

Figure 1: Visual interface architecture and its communication with the other STAR components.

The textual query module, in your turn, verifies textual queries syntax and translates it to the database query language (KOALA language). The STAR data management is supported by the system KRISYS, a knowledge base management system [7]. KRISYS processes a KOALA statment and returns a query result to the textual query module. A data dictionary STAR-KRISYS allows the mapping from STAR objects to KRISYS objects and vice versa. This data dictionary is accessed by the textual query module during the translation to KOALA statements as well as during the query result analysis. This analysis results in a query table (of STAR objects) which is sent to the visual query module to be displayed in the interface.

The visual query language is composed by a window hierarchy, representing the different levels of the object hierarchy in the STAR data model. Several windows can be opened allowing the visualization of many design object representations in the object schemata at the same time. All windows present a group of standard menu options and mouse events, providing a uniform interaction. Some interface operations are exemplified in the following subsections.

3.1 Browsing Facilities

Browsing facilities include navigation through the object schemata, visualization of object properties, version selection, and query activation. Navigation through the object schemata is the most common operation performed by the designer, offering an interactive mechanism to traverse the object's definition following its several representations.

A short browsing session is described in the following to illustrate the interaction with the STAR framework through the visual interface.

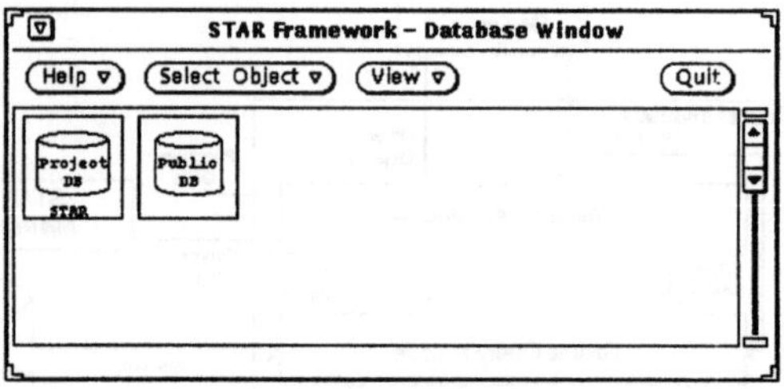

Figure 2: Database window.

The database window is opened when the visual interface is activated. This window displays *Project databases* accessible by the designer as well as the *Public database* (figure 2). The **STAR** project database can be selected (figure 3) leading to a project database window, where the designer's private database, together with two libraries objects called **Risco** and **Calc**, are shown.

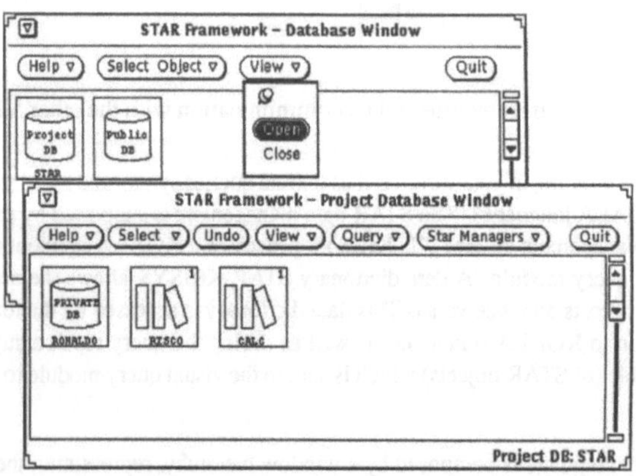

Figure 3: Project database window.

Different icons identify different object types. *Library* icons for example, are displayed as a group of books. When the designer knows the path to a specific object in the hierarchy, it is possible to visualize it by using the *Select-Object* menu option. Figure 4 shows the selection of the *ViewGroup* **Slice** (down to *Design* **Op**) related to the *Library* **Risco**, as described in the object line command.

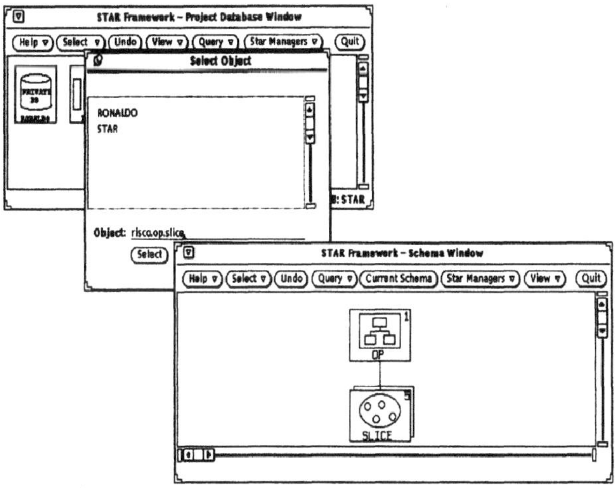

Figure 4: Schema window.

The schema window displays the objects selected. This window allows the visualization of the object schemata from the *Design* node down to *View* nodes.

Visual means are also employed to emphasize object versions. Next and previous versions are denoted by extra icons placed at the bottom-left and upper-right corners respectively (see Slice node in Figure 4). The number associated to each object icon indicates which node version may be currently investigated. Another version of the node can be investigated if its icon is selected with the mouse.

Visualization of object properties is provided by choosing a node with a specific mouse button. This action opens an *object properties window*. The properties of the *Slice* object are shown in Figure 5. Properties that are references to other objects are indicated by its value (the name of the referenced object) followed by an arrow. Once the arrow is selected, another object properties window is opened, showing the properties of the referenced object.

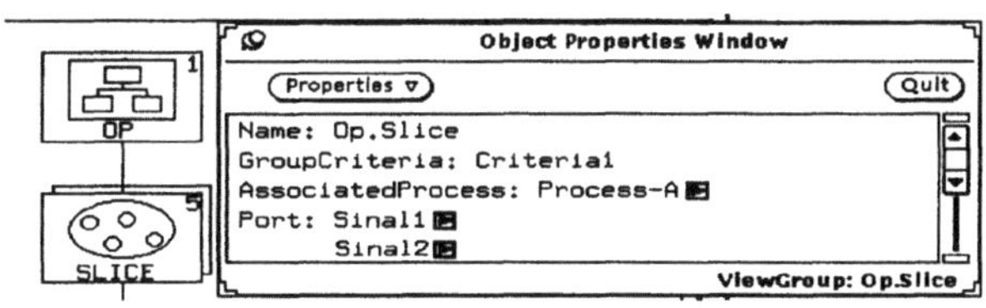

Figure 5: Object properties window.

Investigation of object descendants in the object schemata can be done by selecting an object icon. In figure 6 it is possible to see three *Views* (**LO**, **ENL** and **Extracted**) as the result of the selection on the node **Slice**.

In the same way, if LO node is chosen, a *viewstate window* is opened, showing the *ViewState* called **VsLO1** and its two derived ones: **VsLO2** and **VsLO5** (figure 7).

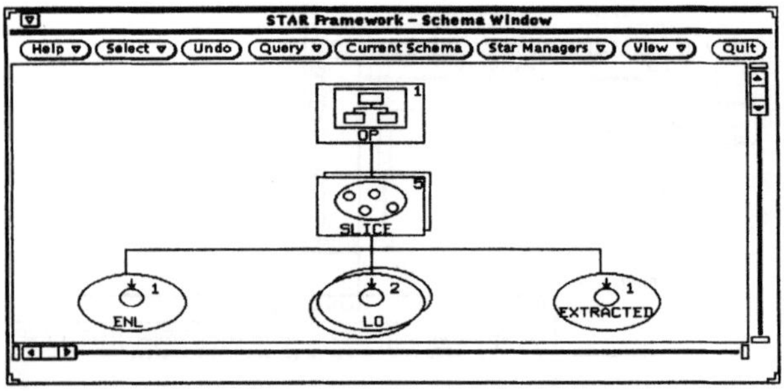

Figure 6: Visualization of **Slice** descendants.

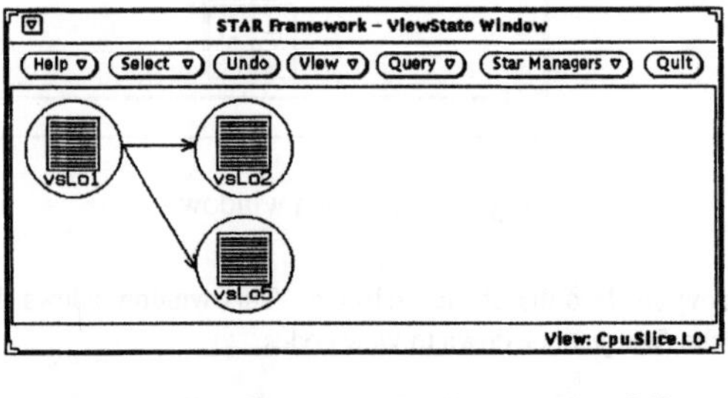

Figure 7: *ViewStates* graph related to *View* **LO**.

3.2 Query Facilities

Queries can be formulated while browsing the database, for example, to restrict the selection of descendants or versions of a specific node. Queries can be defined in a graphical (interactive) or textual way. Through a query menu, which can be opened when a node is selected, the designer can choose the query formulation type.

Graphical queries allow the definition of simple predicates associated to a single object type. When the graphical option is selected from the query menu, the designer initially chooses the kind of relationship, like versions or descendants. In case of descendants, another menu question the designer about the descendant object type intended to be queried. After this, it is still possible to choose between retrieving all or some instances of the selected object type. If the second option is choosen, a graphical query window is opened to help the designer in the formulation of query conditions. An example of the sequence described above can be shown in figure 8, considering that the designer intends to search for some *Views* related to the *ViewGroup* Slice.

Query conditions are defined in an easy way in the graphical query window. Object type properties, conditional symbols and constant values can progressively be chosen to define query conditions that can be combined by logical operators. Figure 8 also exemplifies a condition, displayed at the *Query* line command, that retrieves *Views* which present *Ports* with direction in.

The *Query* line command can optionally be used by the designer to specify conditions in a textual way. Query conditions can be modified, either to update it or to remove it completely.

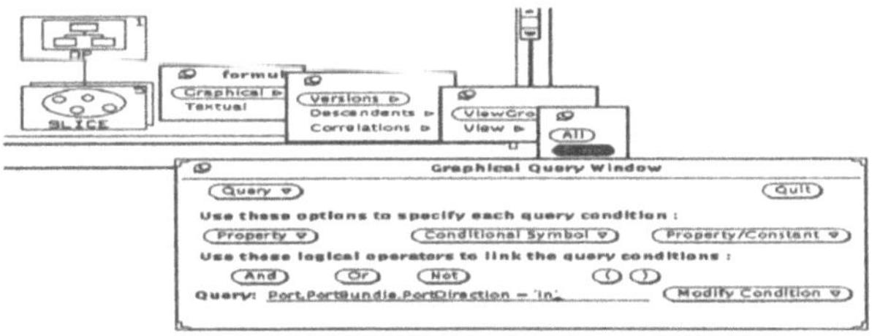

Figure 8: Specifying a graphical query.

Textual queries are defined using a textual query editor. In the next subsection the STAR textual query language is described through some examples.

Graphical and textual query answers can be shown either graphically or textually. In the first case, the objects retrieved by the query are displayed in an iconic form, together with their inheritance and versioning relationships to other objects. In the second case, a *query table window* is opened and the user can navigate on it.

It is possible to save query descriptions and query results after query formulation. Query descriptions can be further loaded in the graphical query window or textual query editor and re-used in new queries. Stored query results can be selected (menu option *Query-SavedResults*) and other queries can be formulated over them to retrieve some of its instances (incremental query facility).

4 The textual query language

Queries can also be formulated by using the textual query language of the STAR framework. The textual query language is an SQL-based language with extended constructs for version manipulation and predicates that deal with complex attributes and relationships in the inheritance hierarchy. Figure 9 shows the basic structure of the language.

select <result_list>
from <object_types> [<version_clauses>]
[where <condition>]

Figure 9: Textual query language structure.

Object versions can be queried by using specific clauses linked to an object variable in the from clause. These clauses are: *firstversion*, *lastversion*, *predversion*, *succversion* and *versions*. Except for the *versions* clause, that retrieves a set of versions, all other clauses return a single node version of the STAR object schemata. The query described in figure 10, for example, returns **Op.Slice** versions which holds the *GroupCriteria* named **Criteria1**.

Some object attributes can be defined as references to a single object or a set of objects in the database, being called complex attributes. In the first case, it is possible to define conditions over the referenced object attributes by using a **dot notation**. Manipulation of set attributes is provided by the

```
select v.*
from viewgroup vg versions v
where vg.Name = 'Op.Slice'
and v.GroupCriteria = 'Criteria1'
```

Figure 10: Using *versions* clause.

clauses *in* and *contains*, which tests element-set and set-set relationships, respectively. The following query (figure 11), looks for *Port* names of the *View* **Op.Slice.ENL** with direction **out**. *Port* is a set attribute of a **View** object type. In addition, these *Views* must be associated to *Process* called **Process1**. *AssociatedProcess* is a reference to a object type *Process*.

```
select p.Name
from view v; port p
where v.Name = 'Op.Slice.ENL'
and p in v.Port
and p.PortDirection = 'out'
and v.AssociatedProcess.Name = 'Process1'
```

Figure 11: A query predicate holding complex attributes.

Queries dedicated for retrieving object descendants or the father of a specific node are formulated by using the clause *is descendant of*. A query example is shown in figure 12. It returns all *ViewGroups* names that are descendants of the *Design* **Op**. The optional character '*' means to search for all *ViewGroups* descendants in the subtree under **Op**.

```
select vg.Name
from design d; viewgroup vg
where d.Name = 'Op'
and vg is descendant * of d
```

Figure 12: A query predicate that tests relationships in the inheritance hierarchy.

Saved query results can also be queried, providing incremental queries facility. Query results names are refered in the from clause preceded by the character '@'. Supposing a query that retrieves data about *ViewStates* was stored with the name **Vs_result**, a new query, described in figure 13, returns *ViewStates* names from **Vs_result** that present the *ViewDescription* **file_ga**.

5 Comparison with other interfaces

Compared to other EDA framework interfaces, the visual interface of the STAR framework offers more facilities related to query formulation.

Nelsis design platform [8] provides meta data and design objects visualization and manipulation. Textual queries can be formulated interactively by selecting object types and attributes in the meta data interface and specific browsers makes possible navegation through different kinds of design object relationships. However, it does not allow graphical queries formulation in the same way of the STAR

```
select vs.Name
from @Vs_result vs
where vs.ViewDescription = 'file_ga'
```

Figure 13: An incremental query.

visual interface, which is completely free from a textual syntax. Visualization of various design objects at the same time, provided by the STAR visual interface, is not an available facility either in Nelsis platform.

CADLAB desktop [9] offers a hierarchical browser to investigate a specific design object. Nodes can be selected and pre-defined methods may be applied to it. New windows can be opened for visualization of children nodes. The browser described by Gedye and Katz [10] allows graphical investigation of composition, versioning and equivalence relationships in a interconnected web of design objects. Nevertheless, query facilities are not supported by them, even in a textual way.

6 Conclusions and future works

This work describes browsing and query facilities supported by the visual query language of the STAR framework. A visual approach to the user interface helps the designer in understanding the complex structure of a design object, avoiding a deep knowledge of the STAR data model.

Most of the facilities available in database visual interfaces are taken into consideration by the STAR visual interface [11, 12, 13, 14]:

- schema visualization and manipulation (browsing);
- interactive (graphical) query formulation;
- textual query formulation for more specialized users;
- incremental queries;
- undo operation while browsing the database;
- facilities for new, as well as specialized users, in activation of operations;
- exhibition of different database objects at the same time (more than one window opened);
- uniformity of menu and mouse options in all windows of the interface;
- objects can be shown in different forms (icons, tables, and so on).

Currently, the visual interface is responsible for managing the browser and the formulation of queries. Communication with the STAR managers, for other design tasks related to data manipulation, is not still implemented.

Other future work concerns graphical queries. Complex queries formulation, which involves, for example, more than one object type or attribute projection in a query result, should be supported in order to increase graphical queries expression power. At present, complex queries can only be formulated in a textual way.

Aknowledgements

We greatfully aknowledge the work of Eduardo A. Rambo who made the implementation of the STAR visual interface windows, as well as the database objects visualization functions.

References

[1] T.J Barnes, D. Harrison, A.R. Newton, and R.L Spickelmier. *Eletronic CAD frameworks*. Kluwer Academic Publishers, 1992.

[2] F.R.Wagner, L.G.Golendziner, and M.R.Fornari. A tightly coupled approach to design and data management. In *EURO-DAC*. 1994.

[3] F.R.Wagner et al. Design version management in the STAR framework. In *3rd IFIP Intern. Workshop on EDA Frameworks*. North-Holland, 1992.

[4] M.R.Fornari, L.G.Golendziner, and F.R.Wagner. Schema evolution in the STAR framework. Also to be presented at *4th IFIP Intern.Working Conf. on EDA Frameworks*, 1994.

[5] F.R.Wagner. L.G.Golendziner, and H.G.Ribeiro. Configuration management in the STAR framework. Also to be presented at *4th IFIP Intern. Working Conf.on EDA Frameworks*, 1994.

[6] R.S.Mello. *A visual query language for the STAR framework*. Porto Alegre, UFRGS, 1994. (Master's thesis, in Portuguese)

[7] N.M.Mattos. *An approach to knowledge base management: requirements, knowledge representation and design issues*. Universität Kaiserslautern, Fachbereich Informatik, 1989. (PhD thesis)

[8] P.Bingley and P.van der Wolf. A design platform for the NELSIS CAD framework. In *27th Design Automation Conference*. ACM/IEEE, 1990.

[9] Cadlab. *Cadlab Framework - Desktop 1.0 User's Guide*. Paderborn, Cadlab, 1991.

[10] D. Gedye and R. Katz. Browsing the chip design database. In *25th Design Automation Conference*. ACM/IEEE, 1988.

[11] H.Kim, H.F. Korth and A. Silberschatz. PICASSO: a graphical query language. *Software Practice and Experience*. v.13, n.3, 1990. p.169-203.

[12] C.T. Wu and D.K. Hsiao. Implementation of visual database interface using an object-oriented language. *Visual Database Systems*. Elsevier, 1989. p.105-125.

[13] M.Leong, S. Sam and D. Narasimhalu. Towards a visual language for an object-oriented multimedia database system. *Visual Database Systems*. Elsevier, 1989. p.465-495.

[14] R. Agrawal, N.H. Gehani and J. Srinivasan. Odeview: the graphical interface to Ode. In *ACM-SIGMOD International Conference on Managment of Data*. ACM/IEEE, 1990.

Glossary

During a discussion meeting in Gramado a couple of participants of EDAF'94 realized that the framework community uses many technical terms having a variety of meanings in mind. This Tower of Babel makes it difficult to compare approaches: do such aproaches solve the same problem or different ones? Therefore it has been decided that EDAF should try to organize a "glossary" in this field. In order to start this process, a first draft of such a glossary is included in this book. All participants have been invited to contribute to this process. The following document is a result of this effort. It combines three major inputs:

1. Nikil Dutt and Ahmed Jerraya compiled the results of the on site discussion at EDAF. This input is labelled (Dutt/Jerraya) in the following glossary.

2. As a reaction to an initiative of Mike Newman, Commission of the European Union and member of the EDAF'94 program committee, Günter Krannich, Siemens Nixdorf Informationssysteme AG contributed the glossary of the JESSI Common Framework Project (JCF), a project supported in part by the commission of the EU under ESPRIT contract 7364. This input is labelled (JCF).

3. Peter Sutton, Department of Electrical and Computer Engineering, Carnegie Mellon University, Pittsburg PA, USA offered his point of view, partially based on literature. His input is labelled (Sutton).

In order to serve as a discussion basis these three inputs have been merged by alphabetical order of the terms to be defined without any comments.

Access Control

(JCF): Access control forms relations between users with different roles, methods, and objects. Access control specifies and enforces constraints to the ability of a users and tools invoked by the user to perform methods on objects. Logical Access Control, Physical Access Control

Accounting

(JCF): Accounting is a collection of values of some system parameters (e.g. CPU time, disc space) for statistics or settlement.

Activity (or CAD activity)

(Sutton): A tool independent design function,
for example:
circuit simulation. An activity is an abstraction of a tool or, for multi-functioned tools, a tool function. There may be many tools that can perform a given activity, and a given tool may be able to perform many activities. An activity can also be called a primitve task.

Activity Model

(Dutt/Jerraya):

- flow of data
- describes process
- high-level, non-structure-related model (behavioral model)

Activity Template

(JCF): An activity template is a subfunction of a tool, determined by a set of parameters. In the framework tool execution is achieved by means of activity template execution.

Alternative

(JCF): An alternative is an exchangeable component, i.e. the component interfaces of alternatives should not differ, but their content: one component might compute faster than the other, but is more expensive.

Archiving

(JCF): Archiving is a form of long-term persistent storage of a complete design with slow access requirements and high durability. Backup.

Audit Trail

(JCF): Audit trails are a form of logging that is automatically generated by a CAD system, or some component of a CAD framework, to keep track of (mis)use of the system by tracking which user used which tool on what data at what time and in what location.

Backup

(JCF): Bakups are periodical saves of the state of the system on permanent storage (i.e.for protection and recovery from system failure and user errors).

Base Communication Module

(JCF): The inter-process communication (IPC) among data base servers and between data base servers and tools is performed by means of a base communication module (BCM). It provides common connection-oriented local and network-wide communication as well as an integrated name service.

Batch Job

(JCF): A set of non-interactive activities to be executed under the control (administration) of the framework at a specified time (in parallel or in sequence) forms a batch job.

Batch Job Support

(JCF): Batch job support is a framework service which allows the definition, execution, and control of batch jobs.

BCM

(JCF): Base Communication Module

Binding

(JCF): If an object has alternatives or versions, relations to that object are not unique. A relation can refer to a special alternative or version (static binding) or to an alternative or version with a certain attribute (dynamic binding).

Black Box Integration

(JCF): This term should not be used any more. Tool Encapsulation

Browser

(JCF): A browser is a service which provides the selection, retrieval, and presentation of information from the framework.

CACE

(JCF): Computer Aided Concurrent Engineering Concurrent Engineering

CAD

(JCF): Computer Aided Design

CASE

(JCF): Computer Aided Software Engineering

CAx

(JCF): CAx means Computer aided engineering where x stands for a given application domain (e.g. CAD or CASE).

Class

(JCF): Object Type

Command Language

(JCF): A command language is a language to perform the interactive desktop operations in a controlled non interactive way. It should be able to create, delete, manipulate, and query the attributes of objects inside the framework.

Composed Relation

(JCF): A composed relation is a relation of two tasks where the design data of the subtask is used for creating design data of the parent task.

Conceptual schema

(Sutton): Same as data schema.

Concurrency Control

(JCF): Concurrency Control is a mechanism to prevent concurrent write access to the same object at the same time. In the framework concurrency control is realised by a workspace concept.

Concurrent Engineering

(JCF): Concurrent engineering is an approach to parallel processing of activities which are needed for the conception, the design, the production and the whole life cycle of a product.

Configurable Framework

(JCF): A configurable framework is a framework which allows custom- or site-specific tailoring of its parts (e.g. the user interface, version management).

Configuration

(JCF): A configuration is a set of related objects that forms a consistent state of development. Objects include task versions and results.

Configuration Management

(JCF): Configuration management is a framework service to define and maintain configurations.

Consistency

(JCF): Objects are consistent if they are correct against their formal specification (data schema). Consistency covers type integrity, referential integrity as well as concurrency control. A system might be considered "consistent" if and only if all assertions (predicates) over the state of the system evaluate to 'true'. Derived data does not alter consistency, no matter at what time it was created.

Customization

(JCF): The process of tailoring the framework components to one's needs. This may be done by setting parameters, by writing rules, by using a special language, by editing resource files, etc. Normally appropriate defaults will be provided by the framework; only small parts of the framework will have to be customized based on site or company specific policies. The framework language covers all customization capabilities.

Data Definition Language

(JCF): With the data definition language it is possible to define or manipulate data schemas as part of the conceptual modelling. The data definition language is a part of the framework language.

Data Manipulation Language

(JCF): A data manipulation language is used to add to, change in, remove or retrieve data from a repository. It is part of the framework language.

Data Model

(Dutt/Jerraya):

- meta-model of a database representation
- domain model, "schema", data representations, structures, etc.
- examples: object-oriented model, ...

Data Schema

(JCF): A data schema contains the object types and the relationships between these object types, thereby describing the structure and properties of a specific application environment.

(Sutton): (Diagrammatic) Representation or model of data.

Dataflow (or dataflow graph)

(Sutton): Set of activities/transformations, connected by a set of arcs that represent data flow.

DDL

(JCF): Data Definition Language

Decomposition

(JCF): tasks and task versions and the relations between task versions within a project are called the project structure or the decomposition of the project.

Derived Data

(JCF): Data that has been derived from other data by an automated calculation is called derived data. Output data of a activity template is said to be logically derived from the consumed input data. Note: The activity template must be reproducible - therefore data produced by an interactive editor is not regarded as (data) derived from a given input. Equivalence

Derived Version

(JCF): A derived version is a version created by deriving an existing version. In the context of (JCF) the term "derived version" indicates a relation of the source and target version in the sense of evolution. Both versions are of the same object type.

Design Data

(JCF): Design Data is the content of results. It describes (a part of) a design in some way. Examples are: documentation, source code, layouts, schematics, etc. Design Data does not include version information, author information, etc. Meta Data

Design Engineer

(JCF): A user who has the role Design Engineer works on tasks and executes activities on tasks which s/he is authorized to work with. Role

Design Environment

(JCF): An integrated design environment is a set of mutually coupled computer-based services which efficiently support the design process in a continuous, user-friendly way. In particular the design environment can be built up using framework services anda domain specific tools.

Design Management

(JCF): Design management is the mechanism for controlling and monitoring a design process, as well as design data and tools. This includes scheduling projects, sequencing tool operations, tracking design status information, user management, tool management, version management, flow management, consistency control and configuration management. It means management of all activities and data that is described by the meta data schema.

Design Methodology

(JCF): The design method defines the design stages, their contents and expected order of execution as well as what criteria must be met for moving between the stages. It dictates what processes must occur during each design stage.

(Sutton): Specific method/approach/procedures to be followed when carrying out a design. A design methodology is used to reduce complexity or enforce consistency in a design. A methodology will often be specified in terms of a flow.

Design Methodology Management

(Sutton): Support for the definition, presentation, execution and control of design methodologies.

Design Process

(Sutton): Act of solving a design problem. A design process may rely on one or more design methodologies.

Design Process Management

(Sutton): Control of, and support for the solution of a given design problem.

Design agent

(Sutton): Active entity (person or tool/process) capable of making a design decision.

Design data management (or data management)

(Sutton): Management of design data through things such as versioning, transaction support (locking).

Design entity

(Sutton): Design tool or data.

Design history

(Sutton): A record of what tasks/activities have been carried out specifying what tools and data were used, who performed the task and when. The design history should be complete enough to enable any specific task to be repeated.

Design trace

(Sutton): Same as design history.

Desktop

(JCF): The desktop is a framework service which represents the user interface to all framework components (e.g. the design management) in a graphical form.

Development State

(JCF): The development progress during the creation, modification, and publishing of design data, resources and configurations is represented by a development states.

DML

(JCF): Data Manipulation Language

Domain

(JCF): A domain is a specialized field, e.g. mechanics, electronics, software engineering.

Domain Neutral Tools

(JCF): Domain neutral tools are services which are not specific to a certain domain.

Domain Specific Data

(JCF): Domain specific data is data that belongs to a particular domain.

Domain Specific Tools

(JCF): A domain specific tool is a tool which belongs to a specific domain.

Dynamic Binding

(JCF): Dynamic binding in design management refers to deferring the selection of a version or an alternative until time of usage.

Encapsulated tool

(Sutton): A tool that is able to work with a framework, due to the presence of a software wrapper or encapsulation that implements communication with that framework.

Encapsulation

(Sutton): (1) Software "wrapper" for a tool (or tools) which implements (or uses) a standard communication interface to a framework to enable the tool to work with (or from) that framework. The wrapper typically treats the tool as a black box and operates by performing transformations on the input and output data to match their format to that of the framework interface.

(2) Act of writing an encapsulation (see definition 1)

Environment

(JCF): An integrated CAX-environment is a set of mutually coupled computer based services which efficiently support the CAX-process in a consistent, (user-friendly) way. In the context of (JCF) it is always an iterative extension or a user defined configuration of the pure framework.

Equivalence

(JCF): A relationship between two or more entities stating that they are identical with respect to some user definable criteria is called equivalence.

Execution State

(JCF): The development progress during the execution of single activities on a certain task is represented by execution states.

Extensibility

(JCF): New services can be added to the framework. This means services belonging to any domain specific environment as well as domain neutral services belonging to the framework itself. It also means that existing services must be extensible. Extensions are defined using the extension language.

Extension Language

(JCF): A language to define and to extend the environment is called extension language. The extension language is part of the framework language.

Flow

(JCF): A flow models the dependencies of intermediate activities building an entire design process. Typically, a flow specifies required sequential and allowed concurrence of the activities. In the context of (JCF) dependencies occur when the output data of one activity template is used as input for a successive activity template or when there is a time dependency. The flow model can be represented by a graph of (possibly parallel branches of) connected nodes representing individual process activities (activities). Design Methodology

(Sutton): A sequence of tasks (NOT tools), typically represented with a directed graph. The arrows in the graph may represent a flow of data OR a flow of control from one task to the next.

Flow Instance

(JCF): The flow instance is an object type which represents the flow belonging to a certain task version.

Flow Instance Version

(JCF): Alternative

Flow Management

(JCF): The framework has to control the order in which activities are performed, to guarantee that a design process proceeds according to the specified flow. No activity template may be performed before all its start-conditions are satisfied. Method, Design Management

(Sutton): Support for the definition, presentation and execution of design flows.

Framework

(Dutt/Jerraya):

- infrastructure for integration

- who is it for? (UNRESOLVED)
  - tool developers?
  - users?
  - for managers?
  - for installers (computer system administrators)

(JCF): A framework is a general computer-based infrastructure which offers services for efficiently building and managing integrated CAX-environments.

(Sutton): Software environment to support (increase the productivity of) users who may be designers, managers, tool developers and/or system administrators. The software environment may consist of a procedural interface, libraries, user interface elements among other things. It may provide services such as design data management, flow management, methodology management, design process management (see definitions below).

Framework Administrator

(JCF): A user who has the role f framework administrator may install tools and register users and teams in the framework.

Framework Dictionary

(JCF): The framework dictionary holds all data that is needed to customize the framework to an environment, i.e. the user defined schema, tool integration and tool encapsulation information, and desktop specific information.

Framework Isle

(JCF): A framework isle is a specific framework instance.

Framework Language

(JCF): The framework language is the lowest language level in the framework. Every action which can be performed by the framework can be expressed in terms of the framework language. E.g. actions defined via a graphical interface are mapped to actions of the framework language. As there are several tasks to be covered by languages within the framework, the framework language may include different dialects. This includes data definition language, data manipulation language, extension language and the command language of the framework.

Framework interface

(Sutton): Method of communicating with a framework, typically procedural.

Global Resources

(JCF): The global resources consist of the objects which are common to one framework isle and which are needed by the user to do her/his work properly. Global resources are i.e. users, teams, roles, tools, activities, flows, and viewtypes.

Gray Box Integration

(JCF): This term should not be used any more. Tool Integration.

Hierarchy

(JCF): A hierarchy describes how a system is (recursively) composed of subsystems. That is, which subsystems are used to compose a system and how they are used. A hierarchy can be represented by a directed acyclic graph, where each node is an object. A root of a hierarchy is an object that is not being instantiated by any other object. A leaf node of a hierarchy is an object that instantiates no other object.

History

(JCF): A history is a record of user activities that have been performed, i.e. a record of transactions on results.

Human Resources

(JCF): Human resources are users, roles and teams. Their description is used to control access to all data and methods of the logical data base.

Information Model

(Dutt/Jerraya):

- conceptual model
  - no specific data structure
  - deals with semantics and not representation
  - examples: EDIF, CFI, ...

Instance

(JCF): An object having the properties of the class is called an instance of that object type.

Instanciation

(JCF): Instanciation is the act of using one result within another composed result, i.e. a result "bicycle" uses two instances of the result "wheel". Hierarchy

Instanciation Information

(JCF): Instanciation information describes the way a result is used, e.g. "the front wheel" of a bicycle.

Integrated framework (or integrated design environment)

(Sutton): Framework that has tools integrated into it. Most commercial frameworks would fit into this category.

Integrated tool

(Sutton): Tool that is able to communicate directly with a framework, without needing an encapsulation or wrapper.

Integration

(Dutt/Jerraya):

- integrated tool can't work if it loses its environment
- integrated object knows its environment

(JCF): Tool Integration

(Sutton): Act of modifying a tool's source code to enable it to work (communicate) with a framework.

Inter-Process Communication

(JCF): Inter-process communication is a mechanism to allow independent computer tasks to communicate at a low level. Inter-Tool Communication

Inter-Tool Communication

(JCF): Inter-tool communication is a high level mechanism to allow tool processes to send each other information (design data or control information) at runtime. Inter-tool communication is based on interprocess communication.

Internal details

(Dutt/Jerraya): Question: what is multiple inheritance for hardware?

Interoperability

(Dutt/Jerraya):

- interoperable tool can work even if its environment is lost
- interoperable object can be developed independent of its environment
- may not mean ability to replace one tool with another

Interoperability

(Sutton): Ability of two (or more) tools or frameworks to work together, typically using a common communication interface.

IPC

(JCF): Inter-Process Communication

ITC

(JCF): Inter-Tool Communication

LAN

(JCF): Local Area Network

LDB

(JCF): Logical Data Base

LDP

(JCF): Logical Data Pool

Local Area Network

(JCF): A local area network is a high speed, reliable data communication channel that links together two or more computer systems to permit data exchange.

Logging

(JCF): Logging is the recording information about actions that are performed, the results of those actions, and situations that occur. Similar to a history mechanism, except logging may be done on a finer granularity (e.g. keep track of actions within a single activity template).

Logical Data Base

(JCF): A collection of all data (meta data and design data) which is managed by the framework forms a f framework isle. In the context of (JCF) the logical data base is composed of several logical data pools. One of these logical data pools holds administrative information (e.g. global resources), each of the other logical data pool holds the data of exactly one project (project specific information).

Logical Data Pool

(JCF): A logical data pool is a collection of data which either holds administrative data (administrative logical data pool) or project specific information (project logical data pool). Logical Data Base.

Meta Data

(JCF): Meta data holds informations about results and their relations.

(Sutton): Data about data, e.g. location, time-stamps, usage information (see design history below).

Meta Data Scheme

(JCF): In a framework the meta data schema describes the organization of the meta data in the logical data base.

Meta-tool

(Sutton): Tool that controls other tools or frameworks.

Method

(JCF): A method is a function or procedure which is associated with an object type. Methods implement services on object types. The framework user selects methods through the desktop.

Methodology

(JCF): While the flow models all possible sequences of development steps from a source design process state to a targeted one, in (JCF) a methodology specifies preferred (e.g. site or designer specific) paths in the flow graph from a source node to a target node (e.g. to reflect a top-down or bottom-up approach). This kind of default subflow selection should not be restrictive rather it should be a changeable guideline for the individual site. Design Methodology.

Methodology Manager

(JCF): A user who has the role methodology manager defines activities and flows.

Monitoring

(JCF): Monitoring is the observation of aspects of system behaviour, e.g. performance and data space.

Name Resolution

(JCF): Name resolutions is a service for mapping names of results, which are meaningful to users, to framework internal identifiers (which are needed for proper framework operation).

Nested Transactions

(JCF): A nested transaction is a transaction started and finished inside another transaction.

Notification

(JCF): Notification is a service for communication events between users and/or tools to notify special events occurring during a flow, e.g. a release of a new library or some part of design data.

Object

(JCF): An object contains some data and a set of operations (methods) that can access that data. An object is an instance of an object type.

Object Type

(JCF): An object type (also known as class) is an abstraction of a part of the application environment. It describes this part in terms of its properties that is attributes and methods. An object having the properties of the object type is called an instance of the object type.

Object-oriented

(Dutt/Jerraya):

- information hiding classes AND inheritance
- objects with behavior (coupling data with operations)
- abstraction vs information hiding:
  - abstraction - conceptual model that ignores details
  - information hiding: external view of objects where you ignore

Parallel Transactions

(JCF): Parallel transactions are transactions started in parallel or concurrent with on relation to each other.

Physical Access Control

(JCF): Physical access control deals with low-level operations and objects, which are meaningful to the implementation of the system, e.g. read- or write-operations on data objects.

Policy

(JCF): Policy is a site specific set of rules which determines what operations are permitted or required under predefined circumstances. Policies are used, for example, in access control, team cooperation, versioning, change propagation, configuration management, etc.

Process

(Sutton): Instance of a tool (and its internal data) existing in a computer's memory.

Product Building Support

(JCF): Product building support is a service that allows the description of dependencies of design data and what the related steps are (rules/actions) and provides for execution of these descriptions. This permits the designer to specify high level tasks that automatically perform complex operations. Examples of such services are 'make' in UNIX or 'dsee' on HP/Apollo.

Project

(JCF): Within a framework installation the different products to be designed are managed as projects. Each project consists of related task versions which are parts of the project.

Project Manager

(JCF): A user who has the role project manager is responsible for a project. S/he creates, deletes and manages projects, and assigns flows and teams to projects.

Project Structure

(JCF): Decomposition

Query

(JCF): A query is a request for the retrieval of a specific set of data from a data base. A query is formulated using the data schema of the data base.

Recovery Mechanism

(JCF): The recovery mechanism either commits or rewinds transactions that were in progress when errors occur, such as unavailable global resources, system crashes or security violations.

Representation

(JCF): Representation is a data model or a data structures used by a tool or a collection of tools for representing design data. A representation is also called view.

Resource (or CAD resource)

(Sutton): Something used during the design process. Tools, design data, libraries, and workstations are all resources.

(JCF): Global Resources

Restore

(JCF): Restore is a service to install a system state which was previously saved by a backup or archiving service.

Results

(JCF): The resuls are the smallest logical units of design data which the design management handles. Result versions are input for and output generated by activities. Internal structures are handled by the tool itself. The design management has no knowledge about that.

Robustness

(JCF): Robustness is a measure of the capability of the framework to remain consistent in case of unexpected circumstances, e.g. tool errors, faulty input, or system / host failure.

Role

(JCF): A role determines the level of authorization for a specific user. This level of authorization is represented by the set of methods which may be invoked by the user on objects, depending on the object type.

Rule

(JCF): A rule states a situation or a condition under which actions (including other rules) must be taken. The specification of actions is part of the rule. Within the framework, rules are often used to customize certain parts of the framework to specific needs (e.g. tailor the version mechanism to enforce a maximum of 6 versions per module). Rules are defined using the framework language.

Schema

(Sutton): Representation or model of something - typically diagrammatic.

Security

(JCF): Security is the protection of information against unauthorized disclosure, alteration, or destruction.

Service

(JCF): A Service is a facility to provide a related set of functions. A service can be implemented as a tool.

Static Binding

(JCF): Static binding in design management refers to deferring the selection of a a specific version or alternative, respectively.

Style Guide

(JCF): A style guide is a detailed guideline how to develop a consistent and easy-to-use user interface.

Task

(JCF): A task is an abstraction of a collection of objects that form an entity. A task holds all the information that logically pertains to a specific part of the design. task versions may be related to each other in a netlike structure. A logical collection of tasks may form a project.

(Sutton): A tool independent design function, consisting of one or more activities, e.g. performance verification (which may require a simulation and then a comparison).

Task Instantiation

(Sutton): The act of linking or binding specific tools and data to a task. For example, the task "circuit simulation" may be instantiated to become "Simulate the circuit new_opamp using HSpice".

Task Schema

(Sutton): (Diagrammatic) Representation of tasks supported in a framework.

Team

(JCF): A team is a group of users who work on a common task (usually in one project) to develop one product.

Tool

(JCF): A tool implements one or more services. From the design management's point of view, two kinds of tools exist, domain-specific tools and domain-neutral tools.

(Sutton): A software "element" (which may be one or more executable images) which performs one or more design functions, for example "HSpice" is a tool.

Tool Communication

(JCF): Inter-Tool Communication

Tool Encapsulation

(JCF): Tool encapsulation is the process of embedding a tool within the framework. By supplying a wrapper of additional code the tool is embedded to the framework, thereby providing access to the tool's functions without modifying the tool's source code. Tool Integration

Tool Execution

(JCF): Tool execution is performed by means of activity template execution. The activity template must be defined within a flow and has to be started within the flow instance for a certain task version.

Tool Integration

(JCF): The term tool integration refers to the process of adding a new tool to the framework in a way that maximizes its effectiveness in the framework. This implies that other tools can determine its presence in the framework, make use of its capabilities, if necessary and make decisions about their actions depending on the results (in the form of the data that it produces) of its use. Integration requires that the tool, its related objects and its behaviour be defined within the framework. Integrated tools share common user interface and design representation/storage

Views

(Dutt/Jerraya):

- DB definition: a collection of information that can be derived from elsewhere.
  - a virtual class
  - corresponds to a set of queries
  - view objects directly refer to base objects
- CAD definition: facet, aspect, presentation of design
- if you have a "complete" model, then DB view = CAD view

INDEX OF CONTRIBUTORS

Altmeyer, J. 66, 233
Asar, P. 89

Baake, U. 199
Bampi, S. 179
Böttger, R. 99
Bredenfeld, A. 243
Buijs, F. 221

Câmara, J. 35
Conradi, P. 189

Director, S.W. 159
Dutt, N. 189

Engel, Y. 99

Fornari, M.R. 45

Golendziner, L.G. 25, 45, 253

Härder, T. 3
Heymann, S. 169
Huss, S. 199

Jacome, M.F. 159
Jones, M.C. 76

Kachel, G. 99
Käfer, W. 221
Kahn, H.J. 57, 211
Kolmschlag, S. 99
Kunzmann, A. 131, 139

Lau, R.Y. 57

Meckenstock, A. 13
Mello, R.S. 253

Nolte, D. 99

Petterson, J. 179

Radeke, E. 99
Ribeiro, H.G. 25
Ritter, N. 3
Rockenberg, A. 169
Rundensteiner, E.A. 76

Sarmento, H. 35, 121
Scholz, G. 109
Schubert, J. 139
Schuermann, B. 66, 233
Schuetze, M. 66, 233
Seepold, R. 131

ten Bosch, O. 149

Unland, R. 13

van der Hoeven, A. 149
van der Wolf, P. 149
Veríssimo, P. 121
Videira, I. 121

Wagner, F.R. 25, 45, 253
Wartchow, R. 179
Westerholz, U. 169
Wilkes, W. 109
Wirth, G. 179

Zimmer, D. 13

MIX
Papier aus verantwortungsvollen Quellen
Paper from responsible sources
FSC® C105338

If you have any concerns about our products,
you can contact us on
ProductSafety@springernature.com

In case Publisher is established outside the EU,
the EU authorized representative is:
Springer Nature Customer Service Center GmbH
Europaplatz 3, 69115 Heidelberg, Germany

Printed by Libri Plureos GmbH
in Hamburg, Germany